PRACTICAL GUIDE TO APPLYING, INSTALLING AND MAINTAINING TRANSFORMERS

By: Alfred Berutti, P.E.

Edited by
Robert B. Morgan,
Senior Editor, *EC&M*

Intertec Electrical Group
Overland Park, KS

Practical Guide To Applying, Installing, and Maintaining Transformers

First Printing, October 1994
Second Printing, September 1998
Third Printing, March 2001

Published by
EC&M Books
Intertec Publishing Corporation
9800 Metcalf Avenue
Overland Park, KS 66212-2215
www.ecmbooks.com

ISBN 0-087288-585-2

INTERTEC PUBLISHING
A PRIMEDIA COMPANY

CONTENTS

PREFACE

THERE ARE FEW items of electrical equipment that are as widely applied in a power distribution system as the transformer. Seemingly they are simple items made up of conductors wound around a core, containing no moving parts (or only a few peripheral items such as gages, fans, or tap changers). They also seem to last forever with only minimal maintenance. The tendency, thus, is to ignore them once they are installed until they give some signs of trouble.

Actually, almost all these "facts" are mistaken. Transformers come in an extraordinary range of types and sizes to satisfy the many tasks they are assigned within the distribution system. Nor are they simple! The theory of transformers is complex, and the care that must be taken in their design and manufacture to meet a client's needs are significant. Nor can they be arbitrarily placed within a distribution system without understanding the NEC rules that apply, and the engineering that is required to achieve adequate performance.

The purpose of this book is to bring together as many of the significant facts that should be known about transformers by persons just entering the electrical power industry. It also will serve as a refresher and source book for those who are experienced in the specification, installation, and maintenance of transformers.

NEC rules applying to transformers are many, and they can be difficult to understand. Only some are covered here. This book, however, is a companion book to another *EC&M* book on the subject of transformers, *Understanding NE Code Rules On Transformers*. That book expands on the many Code rules that apply, and provide examples of how these rules are to be applied in typical systems.

Overland Park, KS
July 1994

Alfred Berutti, PE
Editorial Projects Consultant, EC&M

1

TRANSFORMER BASICS

One of the principal advantages of alternating current is the ease with which its voltage can be changed from one level to another. DC transformation can also be accomplished, but needs sophisticated switching equipment. The AC transformer, on the other hand, takes advantage of the natural build-up and decay of the voltage in an AC system every half-cycle to accomplish its task.

While its operation is deceptively simple, there are may laws of physics that first had to be "discovered" before a practical AC transformer could be built. The materials from which the modern transformers are constructed have also evolved significantly from those used in the past in order achieve the high reliability and efficiency of present-day units.

Before proceeding further, it is important to review the basic concepts involved.

APPLICABLE LAWS OF PHYSICS

Magnetism and its effect on conductors is the basis of modern electrical technology. Exactly what magnetism is has not been conclusively proven; its effects are well known. What is known is that if a bar magnet is broken in half, the two halves will have a north and south pole, exactly like those of the original magnet, but weaker. If one of those halves is again broken in half, the same thing happens. This holds true no matter how many times the divisions are made. The conclusion, thus, is that each of the molecules of a magnetic substance (such as iron) are themselves tiny bar magnets.

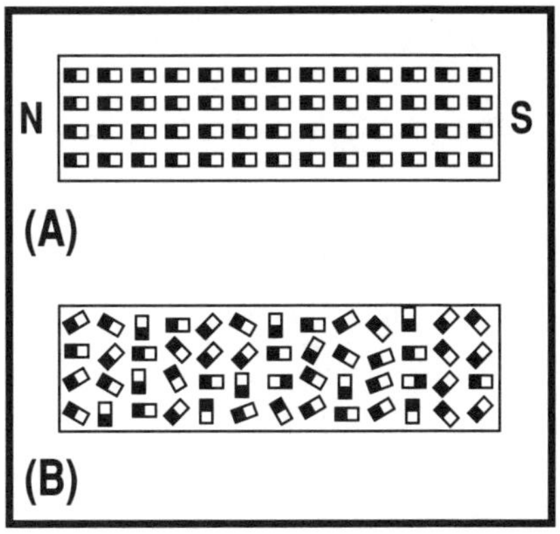

Fig. 1.1. A bar magnet can be thought of as being molecules that are themselves tiny magnets lined up north to south, as in **(A)**, so their magnetism is additive. Heating or striking causes the molecules to assume a random configuration, as in **(B)**, so that their fields cancel each other out and the bar magnet looses its magnetism.

When a bar magnet is pounded repeatedly with a hammer, or heated to a red heat, it will no longer act as a magnet. It can be remagnetized by placing it near another strong magnet. The theory is that in a magnet the molecules are lined up north to south in an orderly fashion, as shown in **Fig. 1.1(A)**. Heating or striking causes the molecules to become randomly oriented as shown in **(B)**, thus tending to cancel each other out.

Many experiments have been carried out to help understand more about magnetism. Many of these are often demonstrated in physics classes in grammar and high schools. Here are some of those that lead to an understanding of the laws that govern the operation of transformers.

MAGNETISM

If a bar magnet is suspended at its midpoint so it can turn freely, as shown in **Fig. 1.2** (next page), it will align itself so that one end will point in a northerly direction. Regardless of how the magnet is turned, that same end will rotate until it comes to rest pointing north. That end is called the *north pole* of the magnet; the opposite side is the *south pole*.

Law of magnetic poles. This law

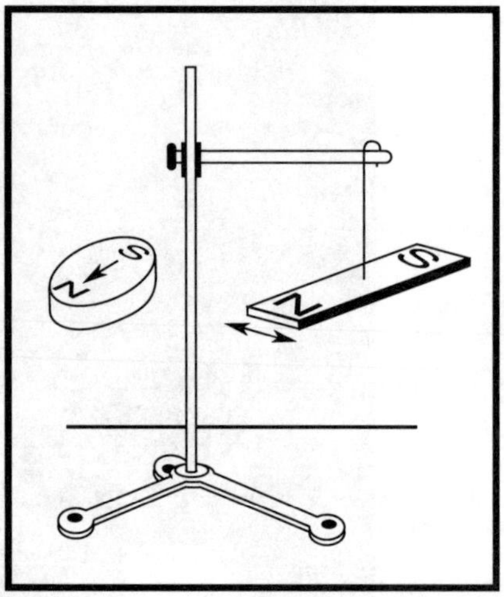

Fig. 1.2. A bar magnet suspended at its middle so that it is free to rotate will always align itself so that one end points in the general direction of north.

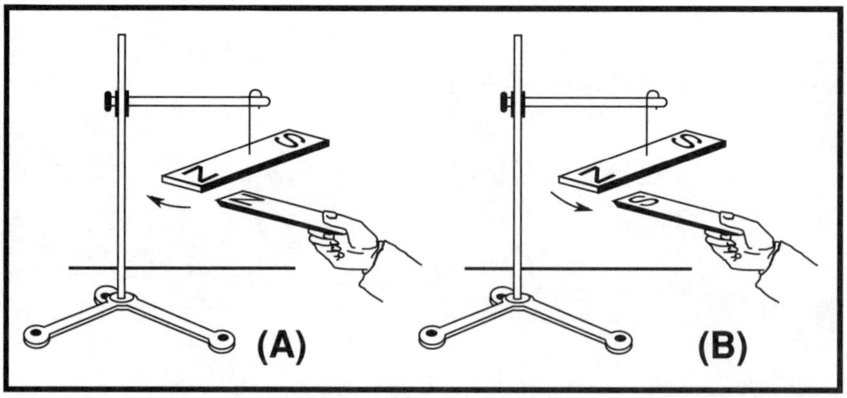

Fig. 1.3. When the north pole of a bar magnet is placed near the north pole of the suspended magnet **(A)**, the suspended magnet will be repelled. If the south pole of a bar magnet is placed near the north pole of the suspended magnet **(B)**, the magnet will be attracted to it.

says that like poles repel; unlike poles attract. This is proved by bringing like and unlike poles of a second magnet near the poles of the magnet suspended at its midpoint. As shown in **Fig. 1.3(A)**, when the north pole of the second magnet is brought close to the north pole of the suspended magnet, it *repels* the suspended magnet. The opposite happens when the south pole of the second magnet is brought near the north pole. As seen in **(B)**, the suspended magnet is *attracted* to the second magnet. Exactly the same repulsion and attraction between like and unlike poles occurs when the second magnet's is brought close to the south pole of the suspended magnet.

The fact that a pole of the suspended magnet will align itself with the unlike pole of the stationary magnet, rather than the Earth's magnetic pole lead to further experiments. It was shown that the force between two magnetic poles is proportional to the strength of the poles, and inversely proportional to the square of the distance between them, or mathematically: $F = S/D^2$. While the strength of the Earth's magnetic north pole is stronger than that of a bar magnet, it is at a considerably greater distance from the pole of the suspended magnet.

Another interesting fact can be deduced. If it is unlike poles that attract, then it must be the *south pole* of the suspended magnet that is being attracted to the Earth's north pole! But by the convention adopted, which is universally accepted, this pole points north and is thus called the *north pole* of the magnet.

The magnetic field. If a magnet is dipped into iron filings, the largest quantity of these filings will adhere at the poles of the magnet, with the fewest attracted towards the center point. The reason can be seen by placing a sheet of cardboard on top of a bar magnet. Now when iron filings are scattered on top of the cardboard, they will arrange themselves to form closed curves as shown in **Fig. 1.4**. These *lines of force* begin at the north pole and end at the south pole. They are much more concentrated at the poles and, thus, more strongly attract the iron filings at the ends. The lines of force repel each other and, thus, tend to spread farther and farther apart as they leave the north pole of the magnet. All these lines of force make up the *magnetic field* of the magnet.

Convention says the lines of force travel from north pole to the south pole. They complete their path through the iron of the magnet. The direction of the lines of force were established experimentally. If a compass is placed at various points along the lines of force, its needle will align itself with the lines of force going towards the south pole of the magnet.

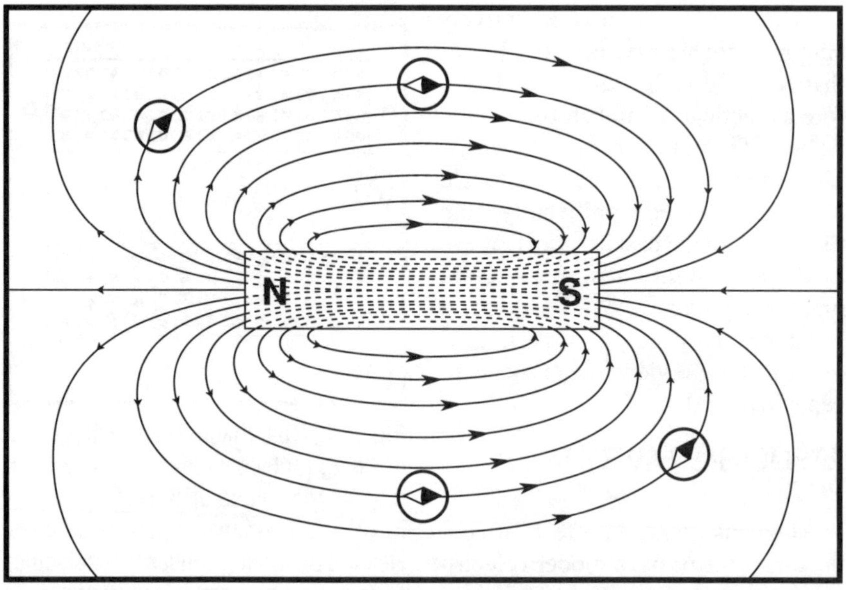

Fig. 1.4. The lines of force (magnetic field) around a bar magnet. This is only a two-dimensional representation. Actually, the lines of force completely surround all sides of the magnet.

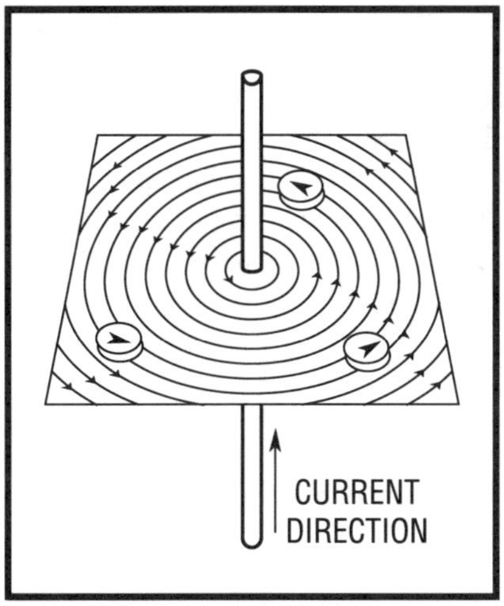

Fig. 1.5. A magnetic field can be shown to exist around a current-carrying conductor.

Magnetic effect of electric current. A magnet is not the only object that is surrounded by a magnetic field. An electric current flowing through a wire also produces a magnetic field around the wire. This is illustrated by the experiment shown in **Fig. 1.5**. A conductor carrying a DC current is passed through a sheet of cardboard and iron filings are sprinkled on cardboard. The filings arrange themselves in concentric circles about the wire. This shows that the magnetic lines of force surrounding the wire are concentric circles, with the wire at the center. By placing a compass at various points in the magnetic field, a direction can be established for the lines of force.

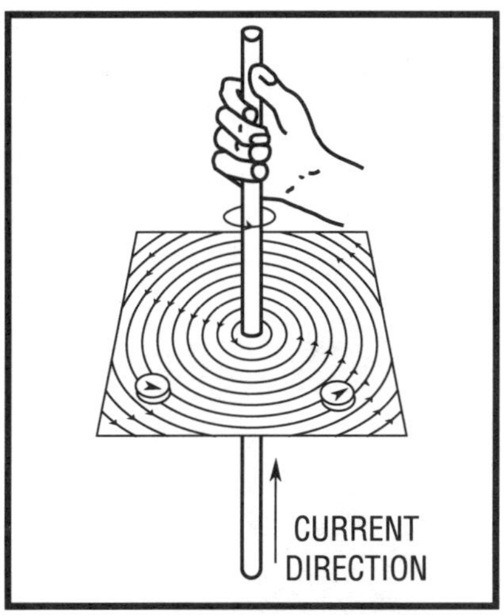

Fig. 1.6. The direction of a magnetic field around a conductor can be determined by wrapping the right hand around the conductor with the thumb pointing in the direction of the current flow. The fingers will point in the direction of the magnetic field.

What can be seen from this experiment is that a fixed relationship exists between the direction of the current in the wire and the direction of the magnetic lines of force. This is illustrated in **Fig. 1.6** and is expressed as follows:

> Using the right hand, grasp the wire with the thumb pointing in the direction of the current. The magnetic lines of force will be in the same direction as the curled fingers.

Magnetic field around a coil of wire. As shown in **Fig. 1.7**, a coil of wire is threaded through a piece of cardboard and a DC current passed through the wire. When iron filings are sprinkled on the cardboards they will align themselves along lines of force that duplicate those surrounding a bar magnet. The direction of the magnetic lines of force are similar to those of a bar magnet; from north to south. What is even more interesting is that strong north and south poles are created that attract opposite polarity magnets and repel those of similar polarity. In effect, the coil of wire acts the same as a bar magnet!

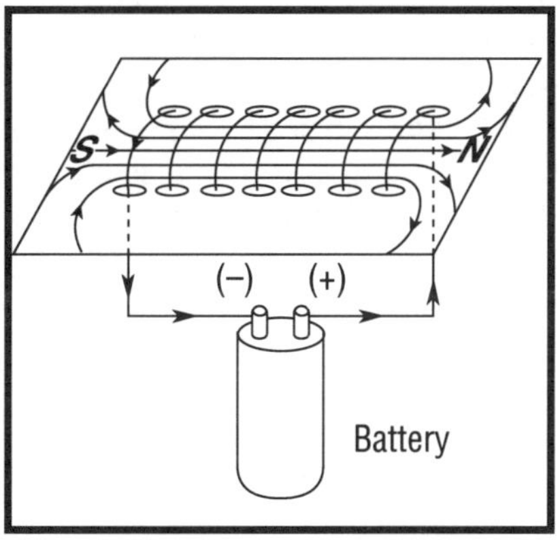

Fig. 1.7. A magnetic field can also be shown to exist around a coil of wire through which a current is flowing.

If a bar of soft iron is placed inside the coil of wire and a current is applied, the iron becomes strongly magnetized. When current is cut off, the iron quickly loses its magnetism. This is the basis of the operation of an electromagnet.

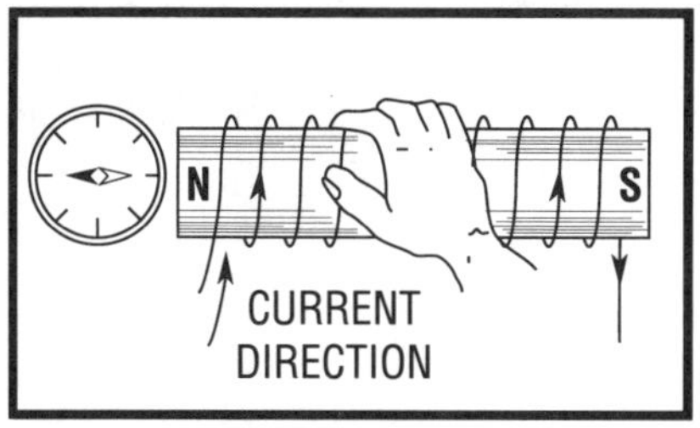

Fig. 1.8. The right-hand rule for locating the north pole of an electromagnet.

The right-hand rule. This is one of the most useful rules for the application of magnetism to electrical circuits. It is simply an application of the results of the previous experiment. As shown in **Fig. 1.8**, this rule makes it possible to determine the north pole of an electromagnet from the direction the wire is wound around the iron bar. It is stated this way:

> Using the right hand, grasp the coil with the fingers curling about the coil in the direction in which the current is flowing. The extended thumb will point in the direction of the north pole of the electromagnet.

ELECTROMAGNETIC INDUCTION

Flow of current through a wire will produce a magnetic field. The reverse is also true. A magnetic field is capable of inducing a current flow through a wire. This discovery is the basis for all electric machinery including transformers.

If a wire is passed rapidly through the magnetic field between the north and south poles of a horseshoe magnet, a current can be detected by a galvanometer. As the wire is passed down (**Fig. 1.9, next page**), the current will flow in the direction shown. When the wire is raised through the magnetic field, the current flow will be in the opposite direction.

If the movement of the wire is stopped within the magnetic field, the current stops. No current will flow if the wire is moved rapidly sideways within the magnetic field rather than up and down. This shows that a current is induced in a conductor only when the conductor is moved so as to

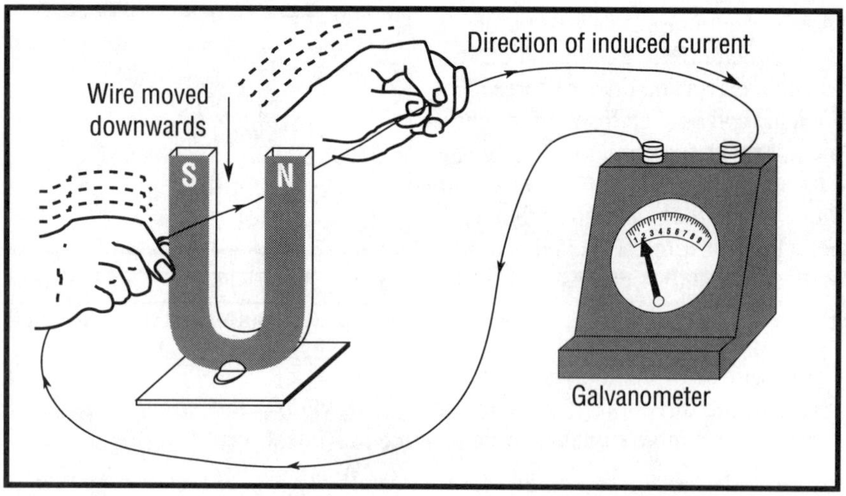

Fig. 1.9. A current can be induced to flow in a wire by moving it quickly up and down between the poles of a horseshoe magnet.

cut the lines of force of a magnetic field.

An electric current cannot flow through a conductor unless there is an electromotive force (emf) driving it. In the experiment shown in Fig. 1.9, it is an induced emf that causes the flow of current in the conductor. The strength of the emf induced in the moving conductor depends upon the number of lines of force it cuts per second; the greater the number, the greater the induced emf. The emf can be increased by:

- moving the conductor at a greater speed;
- providing a stronger magnetic field; or
- winding the conductor into a coil.

Using a coil instead of a straight conductor to cut the lines of force multiplies its effectiveness. A coil with five turns will cut five times the number of lines of force as a single wire moving in the same field at the same speed.

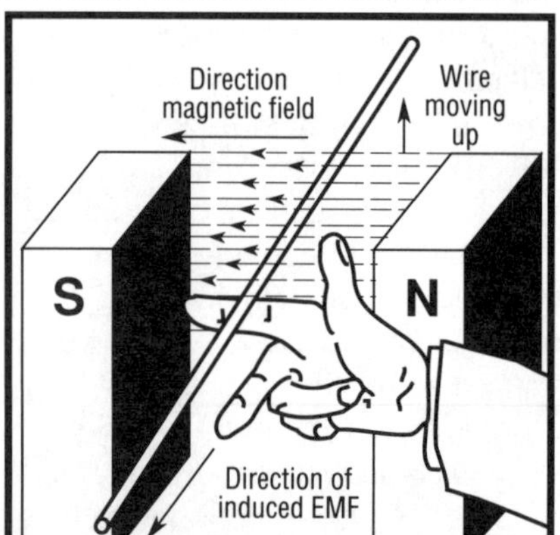

Fig. 1.10. Fleming's rule for finding the direction of an induced emf.

Fleming's rule. It is possible to determine the direction that the induced current will flow by applying this rule as shown in **Fig. 1.10**, and described in this way:

> Using the right hand, hold the thumb, forefinger, and middle finger at right angles to each other. Point the *thumb* in the direction the conductor moves, and the *forefinger* in the direction of the magnetic field (from the north to south pole). The *middle finger* will be pointing in the direction of the induced emf.

When moving the wire down through the magnetic field, the direction of the induced emf (and thus the direction of the current) is in a certain direction. When the wire is moved upwards, the direction of the induced emf is in the opposite direction —an alternating current! This is the operating principle of an AC generator.

Lenz's law. An emf is induced into a conductor when it cuts magnetic lines of force. Exactly the same thing happens when a moving magnetic field cuts across the conductor. This is demonstrated by the experiment shown in **Fig. 1.11**. When the bar magnet is moved quickly through the coil of wire, a current will be detected by the galvanometer. When the bar magnet is stopped, no current will flow. When the bar magnet is withdrawn, a current flow will again be detected, but in the opposite direction.

The direction in which an induced current flows can be determined by applying Lenz's law, which states the following:

> Whenever a current is induced in an electric circuit, it flows in such a direction as to oppose by its magnetic field the motion or change that produced it.

In Fig. 1.11, when the north pole of the magnet is moved through the coil, its magnetic field induces a current to flow in the coil in the direction that creates a magnetic field of its own with the north pole at the front of the coil (see the right-hand rule). Since like poles repel, work must be done to push the magnet through the coil. When the bar magnet is withdrawn, the induced current in the coil will create a north pole at the far end of the coil. Work will again be necessary to pull the north pole of the bar magnet past the north pole of the coil.

Self-induction. As has previously been explained, when a current is passed through a coil, a magnetic field is created that surrounds the coil. When

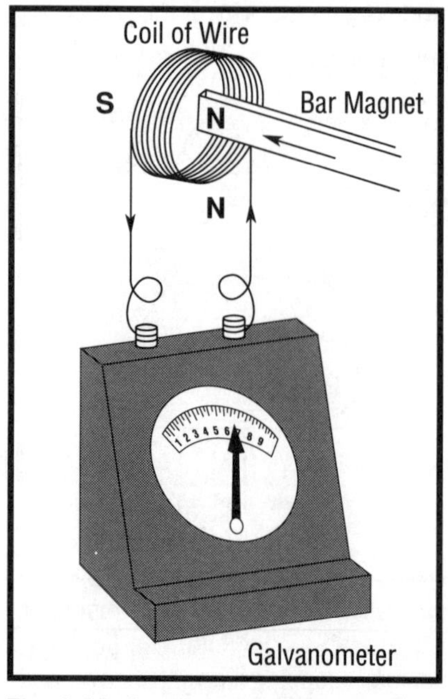

Fig. 1.11. A current can be induced by a moving magnetic field. Lenz's law can be used to determine the polarity of the ends of the coil.

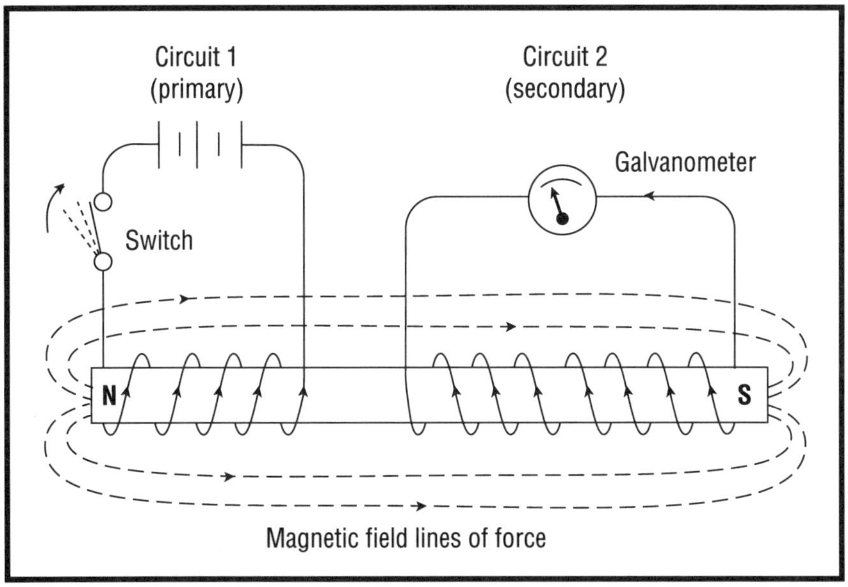

Fig. 1.12. When the switch in circuit 1 is opened or closed, the galvanometer in circuit 2 will briefly deflect and then return to zero.

the current is suddenly interrupted, the magnetic field will collapse. A moving magnetic field cutting through a coil of wire will induce a current to flow within the coil. Thus, the collapse of the field will cause a current to flow that, per Lenz's law, will oppose the change that produced it. Since in this case, it is the stopping of the current that produced the change, the *self-induced* emf will try to keep the current flowing. Arcing at contacts that open a circuit is a good example of this principal.

A BASIC TRANSFORMER

In **Fig. 1.12**, two coils of wire are wound around a soft iron core. When the switch in circuit 1 is closed, current will flow in the circuit in the direction shown. According to the right-hand rule, a north pole will be created at the left, south pole on the right. The soft iron core becomes a strong magnet. Lines of force build up from all parts of the core, cutting outwards through the coils of circuit number 2. According to Lenz's Law, this induces an emf in the secondary circuit that causes a current to flow that builds up its own field, which tends to oppose the buildup of the original field. The needle of the galvanometer in circuit 2 will briefly deflect, then return to zero, indicating that the magnetic field has built up to its maximum.

When the switch is opened, the galvanometer's needle will once again deflect, but in the opposite direction. This is caused by the collapsing magnetic field lines of force cutting through the circuit 2 coils. The current that results from the self-induced emf lasts only as long as the magnetic field takes to dissipate entirely. After that, there is no movement of the field and, thus, no induced emf. This is the basis of all transformers.

An actual transformer is constructed more like the arrangement shown in **Fig. 1.13**. The principles that govern its operation are similar to that described previously, but differ in detail. Here the core is a closed loop rather than a straight bar or rod. The electrical sheet steel used for this purpose is a many thousands of times better conductor of the lines of force (termed the "flux" in transformers) of the magnetic field than is air. The flux, thus, will tend to circulate within the core. This subject will be discussed later in Chapter 2.

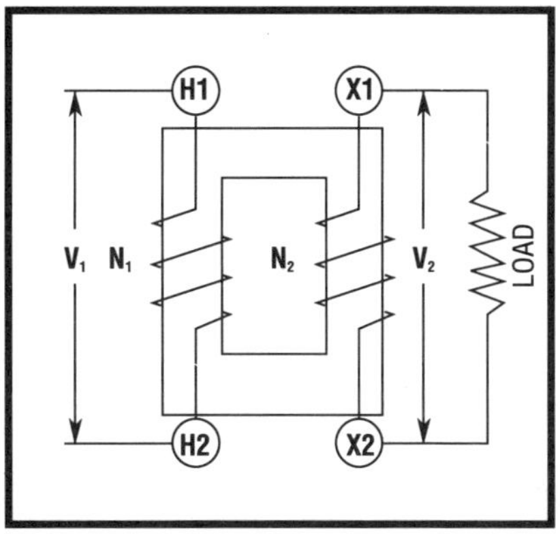

Fig. 1.13. A typical representation of a single-phase transformer.

A transformer can work with a DC source of power, as was described previously. What is needed in these cases is a method of opening and closing the primary circuit as many times per second as needed to achieve the desired AC frequency of the secondary. This, however, is not as convenient as using an AC source in the primary to generate an AC current in the secondary.

AC current flowing in the primary, because it builds up from zero to its maximum plus value and then decays to zero, and continues until it reaches it maximum minus value before again returning to zero, once every cycle, is able to induce a proportional emf in the second circuit on a continuing basis without needing to physically interrupting the current flow in the primary. This is the working principal of an AC transformer.

But a transformer has another significant attribute; it can either raise or lower the voltage in the secondary from the level in the primary. This is the primary purpose of a transformer, although functions such as circuit isolation, generation of a neutral, and others are also part of its value when applied in an electrical system.

A "step-up" transformer has a higher secondary voltage than the primary, and a "step-down" transformer has a lower secondary voltage than the primary. The magnitude of the secondary emf (V_2) depends on:

- the emf (in Volts) applied to the primary winding (V_1) of the transformer;
- the number of turns in the primary (N_1); and
- the number of turns in the secondary winding (N_2).

This is expressed by the following equation:

$$\frac{V_p}{V_s} = \frac{N_p}{N_s}$$

Solving for the secondary voltage, the equation becomes:

$$V_s = V_p \times (N_s/N_p).$$

EXAMPLE: A transformer has 480V AC applied to its primary, which is made up of 100 turns. The secondary of the transformer consists of 25 turns. What is the secondary voltage of the transformer?

ANSWER: Solving for the unknown (V_s):

$$V_s = (480) \times (25 \div 100)$$
$$= 120V.$$

When a transformer steps down the voltage applied to its primary, it increases the current (I) proportionately in the secondary. Thus, in the preceding example, since the ratio of primary to secondary voltage is 4:1, the current in the secondary will be 4 times that of the primary.

Neglecting heat losses, the following relationship holds:

$$V_p \times I_p = V_s \times I_s$$

or

Primary watts = Secondary watts.

This equation says that more electrical energy cannot be obtained from a transformer than what has been put into it.

2

TRANSFORMER TYPES

A transformer is used in so many different applications within an electrical system that it is impossible for one type to be capable of performing all the required tasks. Indeed, while the principle of operation is the same, transformers are probably more varied from one another than almost any other type of electrical equipment.

The basic division of transformers is between power transformers and control and instrument transformers. The first group is enormously varied but all have the function of transforming the voltage within the main power distribution circuits. The second group are usually small units used for metering, to provide power within control and monitoring circuits, or other specialized tasks. Even these divisions are not totally accurate because some transformers, such as grounding transformers, are not specifically used to change one voltage to another level.

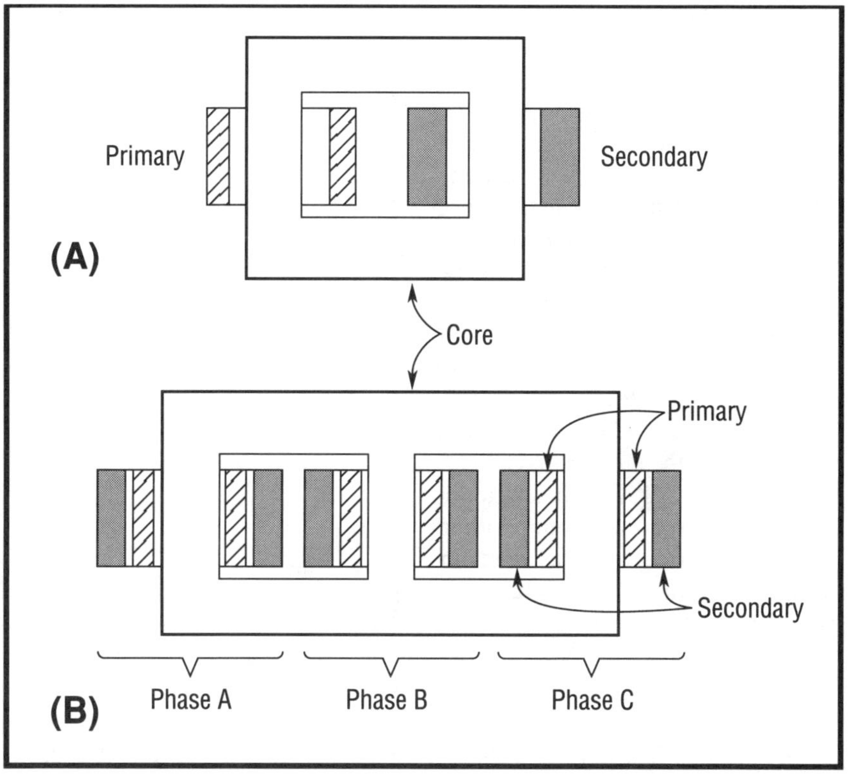

Fig. 2.1. Configuration of windings: in **(A)** the primary and secondary windings of a single-phase transformer are on opposite sides of the core; and in **(B)** the windings of a 3-phase transformer are wound one over the other.

In this chapter the various types will not be discussed in detail, but rather the intent is to provide some general knowledge of subject so that the other chapters in this book can be more clearly understood.

POWER TRANSFORMER CONSTRUCTION

As it was shown in Chapter 1, the existence of mutual flux linking the two windings of a transformer is the key element. As long as the primary and secondary coils are close together, air can act as the medium to carry the flux from one to the other. This, however, will not result in a very efficient transformer because the available flux is diffused over a wide area and a large portion may not cut through the other coil. Wrapping the coils around a common ferromagnetic core provides a better path for the flux to flow because the core material is more permeable than air (has a lower resistance to the flow of flux). A large majority of the flux will, thus, be confined to the core and higher percentage of the flux is certain to link the coils, making for a more efficient transformer.

Ferromagnetic core. In a transformer, the primary and secondary windings can be opposite from each other as shown in **Fig. 2.1(A)**, previous page, for a single-phase transformer, or can be wound one over the other as shown in **(B)** for a 3-phase transformer. Even these two methods of controlling the flux linkages are not totally adequate. While most of the flux will flow through the ferromagnetic material, there still is a small fraction of the flux in the other path (the air). This "leakage flux" does not link the two windings and represents a loss. Further refinements can produce a transformer with even greater efficiency.

Leakage can be limited by different designs of the ferromagnetic path and by subdividing the windings within the transformer. Two designs

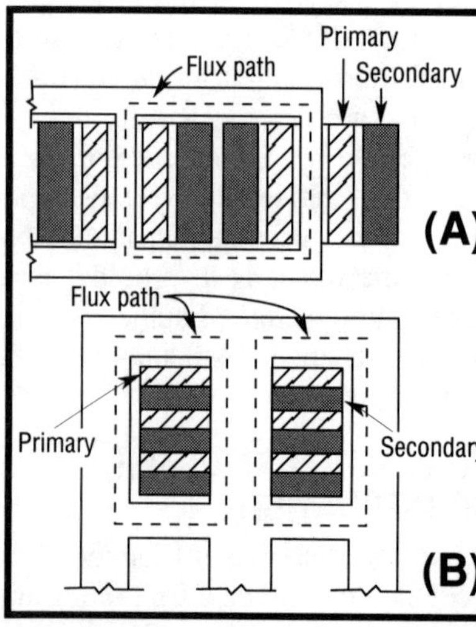

Fig. 2.2. Two different design approaches: **(A)** is a core-type transformer; and **(B)** is a shell-type transformer. In the shell-type, the windings are divided into narrow coils and stacked in pancake style alternating the primary and secondary windings.

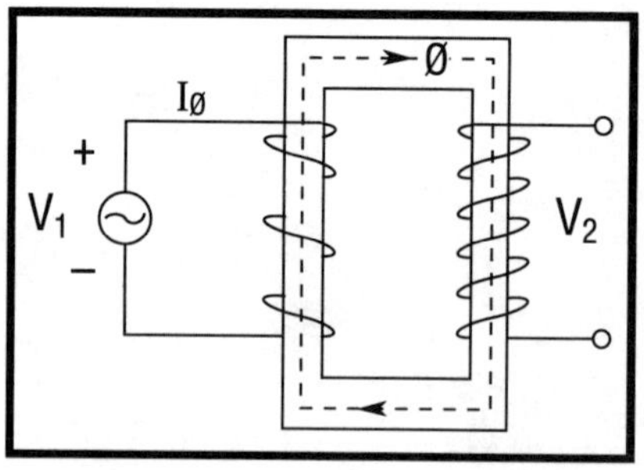

Fig. 2.3. Transformer with an open secondary.

are commonly used for the ferromagnetic structure. As shown for only one phase of a 3-phase unit in **Fig. 2.2**, they are a "core-type (**A**), and a "shell-type" (**B**).

Subdividing the windings is handled differently in each. In the core-type, each winding consists of two sections, one section of each on both legs of the core. In the shell-type, the windings are subdivided into thin pancakes assembled in a stack with primaries and secondaries interleaved. The shell-type construction is used mainly in large transformers where very high short-circuit currents are available and must be resisted. In addition to these two alternatives, there are additional adaptations and techniques that have been used to accomplish the same results.

Load losses. The losses due to the current that flows when a load is connected to the secondary of a transformer is known as the load loss. It consists of the I^2R losses in the windings plus the stray losses due to the leakage flux. Since the stray losses vary approximately as the square of the load current, the total load loss varies just as if it were a true copper loss. In fact, the load loss is often referred to as the *copper loss*.

No-load losses. The illustration in **Fig. 2.3** shows a transformer with its secondary winding open circuited (no-load). When an alternating voltage (V_1) is applied to its primary terminals, a small steady-state current I_θ flows. This is termed the *exciting current*, which in turn establishes an alternating flux (θ) in the magnetic circuit. This alternating flux, in turn, induces a voltage in the secondary (V_2). In Chapter 1 of this book, it was explained that according to Lenz's Law, when the flux cuts through a conductor, a voltage is induced in the conductor, causing a current to flow that will oppose the flow of the magnetic flux in the core. This current is called an "eddy current." It represents a loss in transformer efficiency.

There is another core loss that is due to the way the material that the core is made of reacts to a magnetic field. This is call "hysteresis." Both

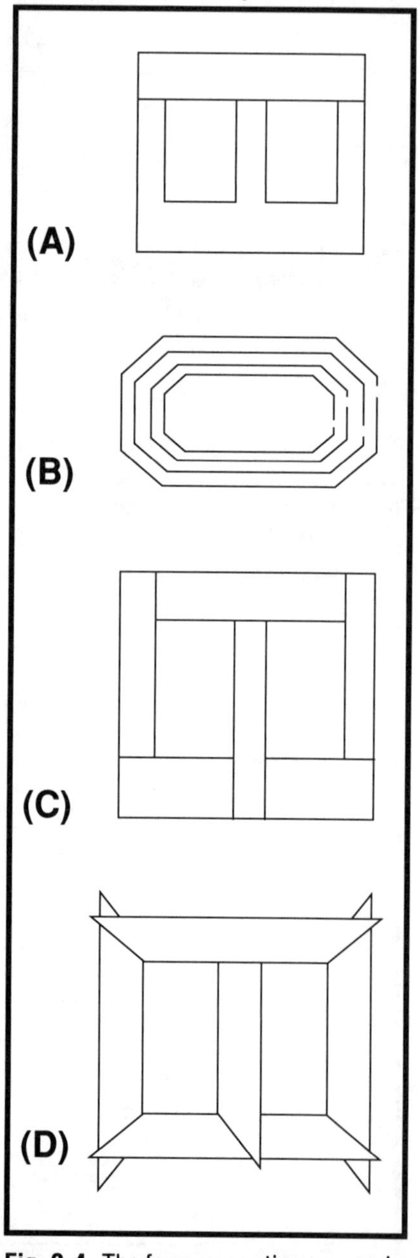

Fig. 2.4. The ferromagnetic core can be assembled from layers of stamped laminations for small transformers **(A)**; wound core, where the core is split so it can be fitted around the windings and then clamped **(B)**; but-and-lap laminations **(C)**; or miter-cut laminations **(D)**.

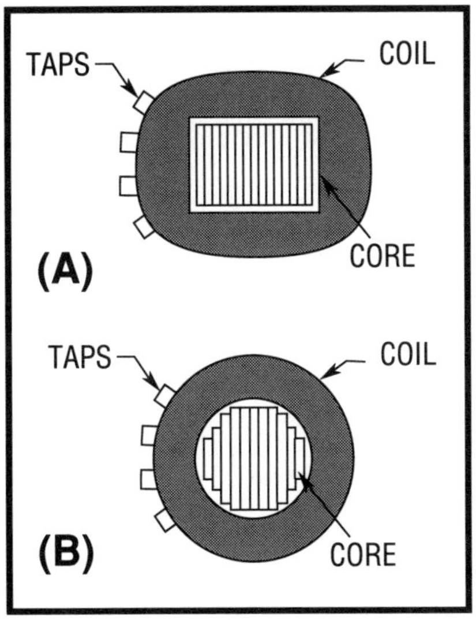

Fig. 2.5. A cross-section through a finished core shows that it can have a rectangular **(A)** or cruciform **(B)** shape.

losses, eddy-current and hysteresis, is minimized in a transformer by the material used and the way the core is constructed.

To minimize these losses, ferromagnetic cores are built up from laminations of coated high-quality silicon steel to maximize the number of parallel paths for the flow of flux. As seen in **Fig. 2.4**, they can either be stamped (**A**) for small VA-rated transformers, form wound (**B**) for smaller kVA-rated transformers, or stacked in either butt-and-lap (**C**) or miter-cut (**D**) construction for larger units. When the proper number of layers are assembled, they are strapped together. When a cross-section is taken through a finished core, the core can be seen to be either a rectangle or cruciform, as shown in **Fig. 2.5**. The type used for a particular transformer depends upon how to achieve the best combination of efficiency, size, and weight.

Economic comparison. The load on a transformer rarely remains constant at its full-load rating. The average load over a period of time is generally lower than that rating. The maximum efficiency, thus, should occur at less than the rated load, depending on the load curve of the intended service. Economic comparison should be based on the efficiency at all loads encountered in the load cycle, and not on just the full-load efficiency. If the transformer is energized continuously, the core loss is present for 24 hours a day; the copper loss is present only when the transformer is delivering useful output. Consequently, the core loss is considerably more important than the full-load copper loss in determining the operating costs of the transformer.

Windings can be either rectangular or round, depending on the shape of the core cross-section, and wound in several different ways.

<u>Barrel wound coils</u> are generally used where layer-to-layer and turn-to-turn voltage stress is low, but high current carrying capability is needed. Insulated magnet wire is wound from one end of the coil to the other, then covered with solid-sheet insulation. Spacers are strapped in place around the periphery and the next layer of coils wound over them. This is repeated until the proper number of coils are in place. The gap between windings allows for air circulation, or if the transformer is to be liquid filled, the circulation of the dielectric fluid.

<u>Continuous disc wound coils</u> are used for units having both high voltage stress and high current. Multiple sections are formed with a rectangular cross-section conductor wound turn-to-turn. Individual columns of insulated spacers are arranged radially to provide support to each turn of conductor. The continuous connection of all sections in series provides excellent impulse distribution.

<u>Sheet windings</u> are often used as the low-voltage coils of transformers. Full width copper or aluminum sheet is wound over a mandrel. The layers of sheets are separated by a paper or fabric insulation, and spacers around the periphery are added where cooling channels are required.

<u>Random wound coils</u> are used where lower voltages and relatively low currents are involved. A round or rectangular cross-section conductor is wound into slotted spacers arranged around the circumference of the coil. The coils are connected in series to provide a high capacitance to distribute impulses through the winding. The problem with this type of construction is in assuring the proper location the conductors in and between each layer is maintained.

Insulation systems. The advantages and disadvantages of dry-type versus cast coil versus liquid-filled units depend upon the application. Dry-type and cast coil transformers usually can be located closer to the load, resulting in cost savings because of shorter cable runs and lower electrical losses. In some applications, however, there is no option to the use of liquid-filled transformers. Dry-types and cast-coil units are limited in kVA size and the types of environments they can be exposed to. Liquid-filled transformers have neither of these limitations. When installed indoors, on the other hand, liquid-filled units may be required to be installed in a special room for fire safety reasons. This may dictate a location some distance from the load.

<u>Dry-type transformers</u> are available for either indoor or outdoor installation. They can operate indoors in normal ambient conditions and are best suited for normal duty where load demands are not severe. For outdoor operations, dry-type transformers are supplied with a NEMA 3R enclosure having louvers for ventilation.

To keep out moisture, windings in a dry-type transformer are usually dipped in varnish at atmospheric pressure, or varnish impregnated under vacuum to assure maximum penetration.

Dry-type power transformers are limited in the size they are available in. Depending on the supplier, they are available up to 2500, 5000, or 10,000kVA.

<u>Cast-coil transformers.</u> Where the windings of dry-type transformers can be adversely affected by hostile environments (dirt, moisture, corrosive fumes, etc.), a cast-coil dry-type transformer is recommended. The space between the turns and between the coils of these units is completely filled with glass fiber sheet and tape (except for the air cooling channels). When completed, the winding is placed into a mold into which an epoxy liquid is introduced under pressure. The glass fiber becomes saturated, and when cured, the entire winding is a solid

mass. The resulting transformer is immune to virtually all atmospheres.

Similar to dry-type transformers, cast-coil transformers are limited in the kVA ratings in which they are available.

Sealed-coil transformers share some of the features of the dry-type and cast-coil transformer. Here, the top and bottom edges of the varnish-impregnated windings are permanently sealed to prevent penetration of moisture. The units can pass a 24-hr submersion test. As with the other types of dry-type and cast-coil units, their kVA ratings are limited.

Liquid-filled transformers. As the name implies, the winding are installed within a tank that is then filled with a dielectric fluid, that can be either oil, silicone, or high-molecular-weight hydrocarbon. The fluid serves as part of the insulation system of the windings as well as a cooling medium. Natural circulation of the liquid is set up inside the tank by the temperature differences. This carries heat away from the windings and transfers it to the tank wall or to external radiators.

The windings, thus, are totally unaffected by the surrounding atmosphere, but are subject of deterioration of the kraft-paper insulation and breakdown or contamination of the insulating fluid. Aramid paper and spacers can be used in place of kraft paper and pressboard spacers to limit insulation deterioration.

Liquid-filled types are available in almost limitless kVA and voltage ratings and can be applied in any type of environment. They are extremely rugged and, if properly maintained, long lived.

Connections. Power transformers can be further subdivided into single-phase and 3-phase units. As the name implies, and as shown in **Fig. 2.6**, the single-phase transformer taps only two phases of a 3-phase or single-phase 3-wire power system (or one phase and the neutral, as indicated by the dotted line). The 3-phase transformer utilizes all three phases (the connection shown is for a delta primary, grounded wye secondary). The different connections are discussed further in Chapter 3 of this book.

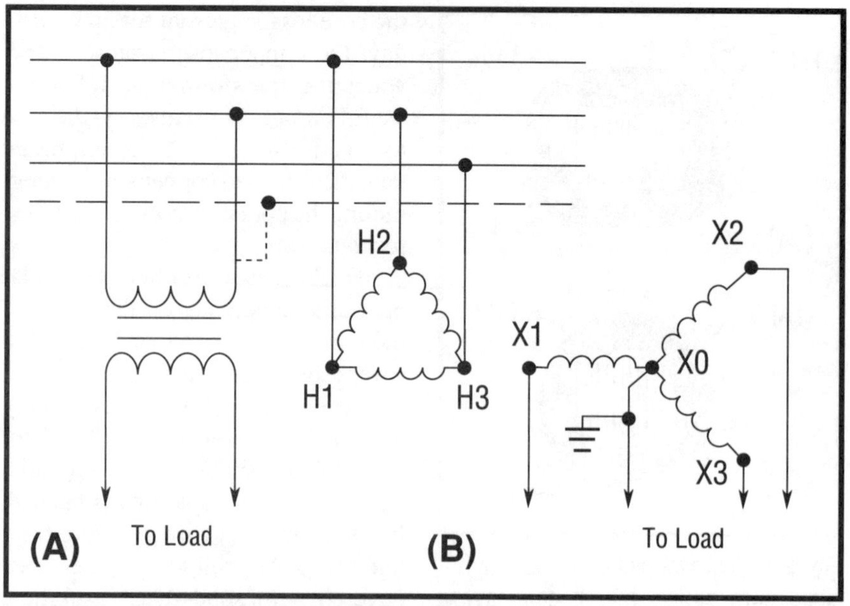

Fig. 2.6. Connections of a single-phase transformer **(A)**; and a delta-wye 3-phase transformer **(B)**. The single-phase unit can be connected line-to-line as shown, or line-to-neutral as indicated by the dotted line.

A great majority of the large transformers within a power distribution system are 3-phase units in which all the phases are contained within a single tank or enclosure. Three single-phase transformers can also be connected together to provide the three phases, but this technique is less frequently used now than in the past except in smaller sizes and to form specialized connections.

SHIELDED ISOLATION TRANSFORMERS

Transformers are very versatile. A dry-type transformer can be fabricated with its windings separated by a grounded metal-foil shield. They are then referred to a "shielded isolation transformers," whose purpose is to reduce noise in power lines that will interfere with the operation of sensitive electronic equipment such as computers, instrumentation, and the like. More information about noise and noise rejection can be found in *EC&M's* book entitled *Practical Guide to Quality Power for Sensitive Electronic Equipment*.

Shielded isolation transformers can be 1:1 ratio, for isolation only, or can also provide voltage transformation. The minimum shielding (**Fig. 2.7**) is usually a single grounded Faraday shield between windings, but additional attenuation can be provided by also shielding either or both individual windings. The amount of noise reduction (attenuation) achieved (Vti/Vto) is expressed either as a ratio or in decibels (db). The table in **Fig. 2.8** shows the relationship of ratio to db.

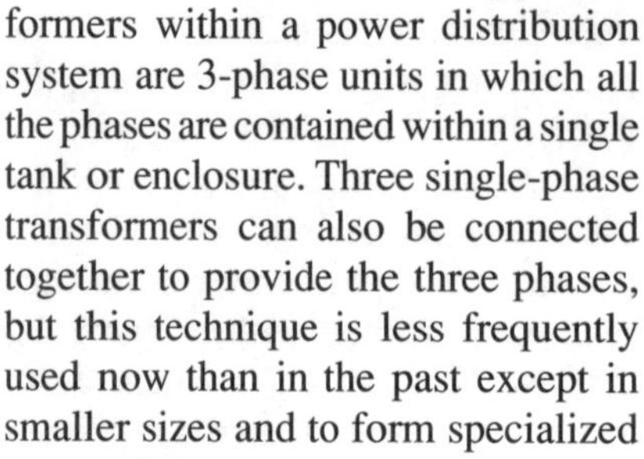

Fig. 2.7. A basic shielded isolation transformer has a grounded metal foil inserted between the primary and secondary windings.

V_{ti}	V_{to}	Ratio	Common Mode noise attenuation (db)
5	1	5:1	14
20	1	20:1	26
50	1	50:1	34
100	1	100:1	40
150	1	150:1	44
200	1	200:1	46
250	1	250:1	48
300	1	300:1	50

Fig. 2.8. Common mode noise attenuation can be described either as a ratio or in decibels (db).

In a shielded isolation transformer, capacitance coupling between the primary and secondary windings of an unshielded transformer is replaced by the primary-to-shield (Cps) capacitance, and that of the shield-to-secondary (Css) that results from the small electrostatic field around the shield. The primary-to-shield capacitance conducts most of the incoming transient energy (Vti) to ground. Thus, the effective capacitance (Ct) between primary and secondary windings of a shielded isolation transformer is much less than that of an unshielded unit

Electrostatic shielding has been routinely used in large DC rectifier transformers with medium-voltage (MV) primaries. Here, shielding protects the secondary thyristors from noise that can cause random firing and failures. So shielding is available and is effective in higher-voltage transformers.

Use of electrostatic shielding on AC medium-voltage transformers has not extensively been employed. There is an advantage in its use on the MV level. The presence of electrostatic shielding on the higher-voltage transformer may mean that other shielded transformers are not required downstream in the system. Even if transformers connected to the secondary of the main transformer also have shields, there is still an advantage to using this technique. When a shielded transformer feeds another shielded transformer, the total attenuation will be one multiplied by the other.

AUTOTRANSFORMERS

Another variation on the standard power transformer is the autotransformer. Here, instead of insulating the primary and secondary windings of the transformer from each other, the two are connected. By doing so, the operating voltage of the system can be raised to a higher level or can be lowered. The resulting voltage can be used to feed equipment needing a higher or lower voltage than is available at the site.

A typical single phase dry-type transformer with two primary and two secondary windings is shown in **Fig. 2.9(A)**. When used as a lower the voltage, say from 480V to 120V, terminals H2 and H3 are connected together to form the primary winding, and X2 and X3 are connected together to form the secondary. The 480V line voltage is then connected to H1 and H4, and the voltage between terminals X1 and X4 is 240V, which is center tapped to derive the 120V between either leg and the neutral.

The same transformer connected as an autotransformer to boost the voltage to 600V is shown in **(B)**. The obvious difference is that the H1 and X4 terminals have been connected together, and the two secondary windings have been paralleled. This type of connection is more commonly illustrated as shown in **(C)**. This is often further simplified as shown in **(D)**. This type of autotransformer can also be used to drop the voltage from 480V to 380V by connecting the higher voltage to X4 and H1, and obtaining the lower voltage from terminals H1 and H4.

A point of interest about autotransformers is that they can be applied to a much larger load than is permitted for their rating when they are connected as a standard transformer. For instance, the autotransformer shown in Fig. 2.8(C) has an output kVA rating of five times its nameplate rating. More information on this is given in the example in Chapter 6 of this book.

Note that NEC requires that an insulated transformer can be reconnected for use as an autotransformer to produce a higher-than-rated secondary voltage only if the transformer is so labeled by the manufacturer. While they may look like ordinary 2-winding transformers, the secondary windings have sufficient electrical insulation to allow them to be connected to the same supply lines as the primary windings.

A "buck-boost transformer" is a special adaptation of the autotransformer limited to making only a relatively small change in output voltage. The secondary windings, thus, are rated at 12 and 24V; 16 and 32V; and 24 and 48V. To boost the voltage from 208V to 240V the transformer with a 16 x 32V secondary would be selected. A further discussion of connections can be found in Chapter 3 of this book.

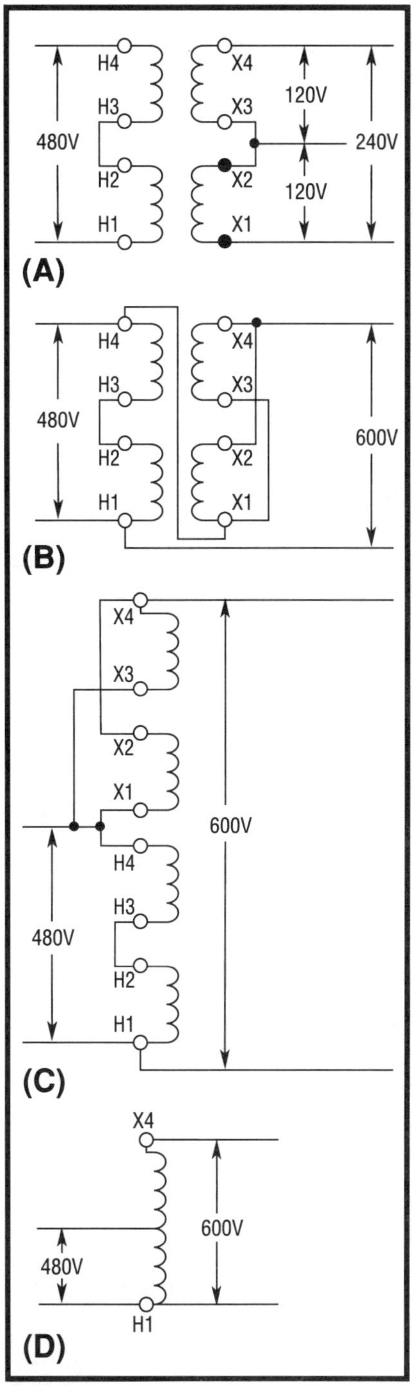

Fig. 2.9. Autotransformer connection. **(A)** shows the connections of a typical single-phase transformer having four windings. The same transformer connected as an autotransformer is shown in **(B)**; this can be redrawn as shown in **(C)**; and further simplified as shown in **(D)**.

A "grounding autotransformer" is another specialized form of autotransformer. Its principal function is to derive a neutral from an ungrounded power source. The neutral is then used in several different ways. The transformers are zig-zag connected or T-connected types installed in 3-phase, 3-wire ungrounded systems to create a 3-phase, 4-wire distribution system, or to provide a neutral reference for grounding purposes. This connection is also discussed in more detail in Chapter 3 of this book.

CONTROL-CIRCUIT TRANSFORMERS

There are a number of designations given to transformers that can be considered as insulating, dry-type, and of a kVA rating (usually 5kVA and smaller) normally associated with control circuits. There are, however, significant differences among them and only some are applicable to specific types of control circuits. The term "control-circuit transformer" is generic and applies to no specific type.

In most cases, all small transformers are capable of generating a lower voltage, such as 12, 24, or 120V from a higher voltage such as 240, 480 or 600V. The primary object of using a transformer for control circuits is safety. A totally new insulated electrical circuit is derived by the secondary of a transformer. Potentials of 120V and below are used to reduce the electrical hazard.

A general-purpose transformer with a wiring compartment.

The following discussion will help differentiate among the designations used to describe the various transformers.

General-purpose transformers are usually rated at 600V and below. They are used for reducing voltages to 120 and 240V (or the many other voltage combinations available), and are used to power light loads such as lighting and fractional-hp motors. They are usually equipped with an enclosure and a wiring compartment. General-purpose transformers are usually voltage compensated such that at full load the output voltage is reasonably close to the nameplate voltage. This group of transformers is not intended to have good regulation (the difference between the output voltage at no load and at full load) since they are not meant to operate momentary loads greater than their nameplate rating.

Control transformers are intended for applications such as supplying pilot light and signaling circuitry where voltage regulation is not critical. Typical secondary voltages are 6, 12, 24, 120, and 240V. In most cases they are made in sizes lower than 250VA (0.025kVA). Typically, they are equipped with an enclosure, but rarely with a wiring chamber.

The secondary output voltage is compensated such that when nameplate voltage is applied to the primary, the secondary voltage will be approximately the nameplate voltage when delivering rated load at 100% power factor. Regulation is not exceptionally good, since this type of transformer also is not intended for applications where the momentary overload is greater than the continuous load rating shown on the nameplate.

Machine-tool transformers are intended to operate relays, contactors, motor starters, and other electromagnetic devices that require very large momentary inrush current. Typically, they have secondary voltages of 24, 120, or 240V and range in size from 50 to 5000VA (5kVA).

An industrial control transformer found in an MCC and other control circuits is capable of sustaining the voltage even during inrush caused by motor starting. This type of transformer is also known as a machine-tool transformer.

Their regulation is usually good, even at extreme momentary overloading such as 6 to 8 times the nameplate rating. *Industrial control transformers* are actually machine-tool transformers.

Class 2 transformers are suitable for meeting the requirements of NEC Article 725 for supplying power to Class-2 circuits. The Code limitation on size is 100VA. Typically, they are used for operating bells, furnace controls, signal circuits, and other low-energy apparatus. Usually, the output no-load voltage is 30V or less.

Class 2 transformers come in two categories: energy limiting and nonenergy limiting. Energy limiting means that when the transformer is short-circuited, it will either limit the amount of current by inherent design, or the transformer is required to have an overcurrent protective device, such as a fuse, that will operate automatically so the unit will not reach an abnormally high temperature when overloaded.

These transformers are generally provided with an enclosure of noncombustible material. The output voltage regulation is usually very poor.

Signaling transformers are often used to supply signaling systems not covered by the Class 2 circuit limitations of the NEC. Typical output voltages are 4, 8, 12, 16, 20, and 24V. Usually these multiple-output voltages are obtained by taps on the same secondary windings. Consequently, the VA capacity is reduced in the same proportion as the secondary voltage. Usually they are available in sizes of 1kVA and smaller. They do not have unusually good regulation characteristics when a load greater than the nameplate rating is placed on the transformer.

VARIABLE VOLTAGE TRANSFORMERS

A number of standard transformers can have, to a limited degree, their primary/secondary ratio changed by using taps located usually on the primary windings. To obtain a greater degree of flexibility in changing the ratio between the primary and secondary coils to allow greater secondary voltage changes to occur, a variable voltage transformer is used. The amount of voltage change depends upon the type of variable transformer used. One main reason for changing the secondary voltage is to provide for compensation when the incoming line voltage changes. Thus, by using automatic equipment, when the line voltage fluctuates, the secondary voltage that serves the load remains constant.

Brush type. To achieve variable output voltage an autotransformer, but of a different type, is used. Instead of having the secondary voltage based upon fixed output taps, one tap is fixed and the other tap is connected to a brush that slides along an uninsulated section of the coil. One way of doing this is to have the coil wrapped around a toroidal shaped core.

The voltage ratio relates to the location of the brush as it rides against the coil and depends upon where on the coil the brush is allowed make contact with. Limited ratio variable transformers are made as well as full range units that can change the voltage from 0 to about 120% of the incoming line voltage. When the output voltage exceeds the input voltage it means that there are extra turns on the coil that extend beyond the windings that lie between the incoming power terminals (in affect the unit becomes a step-up transformer).

A typical brush-type variable voltage transformer.

Ratings start at less than 1kVA for a 120V, single-phase variable transformer. Basic units are wired in parallel and/or in series to obtain higher power capacity. Two units in parallel have double the current and kVA rating. The individual units are stacked on top of each other, bolted together and operated from a common shaft that rotates the brushes. Operating in a configuration that combines parallel and series connections, with several units stacked together, this type variable voltage transformer, at 480V, can have a rating exceeding 200kVA. Sometimes a control mechanism is attached to the rotor that turns the brushes allowing automatic adjustment of the output voltage.

An important characteristic of these type transformers relates to brush contact and the amount of current flowing through the carbon brush. A rating based solely on output kVA may cause serious problems because, for a given kVA load, the current drawn depends on the output voltage. Since the output voltage is variable, a load of a given kVA value may draw a safe current at 100% voltage, but at 25% voltage, the current required to serve a load of the same kVA value would require four times as much current which may cause the brush to overheat. Therefore it is important that the maximum current rating not be exceeded.

Induction type. Another form of variable voltage transformer is the induction type (**Fig. 2.10**) that does not use brushes. The usual voltage change for these units is ± 10% but can be greater. This unit is a variable-ratio autotransformer that uses two separate windings, a primary and a secondary. There is a laminated steel stator on which is wound a winding that serves as a secondary coil. This winding is connected in series with the load. The primary coil is connected across the supply line. This shunt winding is wound around a rotor. Construction is similar to that of a motor except that in this case, the rotor can only turn 180 mechanical and electrical degrees.

As the primary core is rotated, the

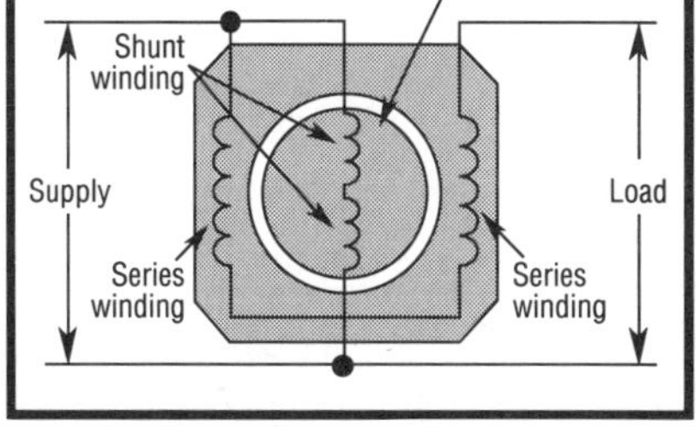

Fig. 2.10. A typical induction-type variable voltage transformer.

amount of primary flux passing through the secondary winding is decreased until the core reaches a position at right angles to the secondary winding. In this position no primary flux passes through the secondary windings and the induced voltage in this coil is zero. The continued rotation of the core in the same direction again increases the amount of flux passing through the secondary, but it is now in the opposite direction and so reverses the direction of the induced voltage. Thus the output voltage can be varied by adding or subtracting from it the voltage induced in the secondary winding.

Both single and 3-phase transformers are available. Ratings of these types of transformers vary from 8kVA, 120V, single-phase to 1500kVA, 480V, 3-phase.

INSTRUMENT TRANSFORMERS

These types are applied to transform the primary voltage or current to a secondary value that is suitable for use with meters, relays, or other measuring devices rated usually at 120V or 5A.

When used as part of a measurement circuit, the transformers must not introduce any unknown error into the system. Their accuracy must, therefore, be either a known value so that the error can be compensated for, or be so negligible that the error can safely be disregarded. ANSI standards define the accuracy of instrument transformers in terms of the maximum amount of error it will introduce. There are three levels of "accuracy class:" 0.3, 0.6, and 1.2. These numbers represent the maximum error (in percent) they will introduce. For example, 0.3 means an error of ±0.3% ($^3/_{10}$ of 1%).

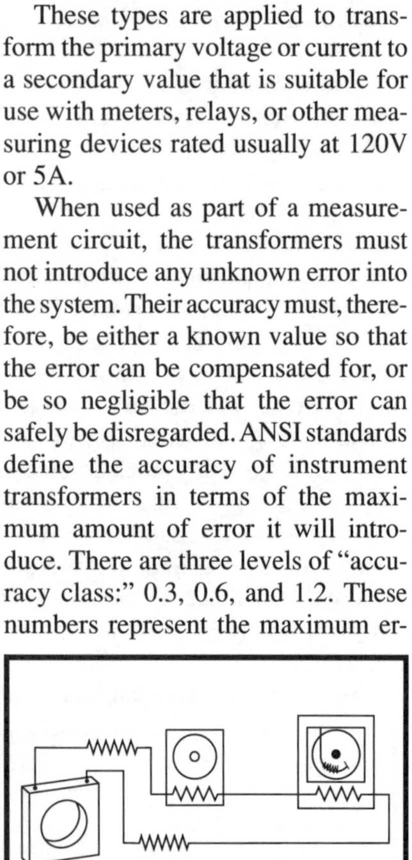

Fig. 2.11. Burden of a CT or PT includes the impedance of the instruments and conductors within the circuit fed by the transformer.

For any specific transformer and circuit condition, the accuracy depends on the *burden* imposed on the secondary of the transformer. The burden is equal to the impedance (ohms) of the instrument(s) it is attached to, plus that of the wiring of the circuit, as shown in **Fig. 2.11**.

Potential transformers (PTs) basically are two-winding, single-phase units specialized for supplying voltage to instruments, metering and monitoring devices. They also provide power for such things as strip heaters in switchgear enclosures, and the like.

Because the accuracy of PTs are dependent on the burden, standard burdens have been designated, and these are the burdens at which the accuracy is classified. The burden of an actual circuit rarely is the same as any one of the standard burdens, but the standard burden serves only as a standardized reference point for stating the accuracy of the transformer.

The accuracy of PTs is affected only slightly by changes in temperature, waveform, and voltage variation. An increase in frequency above the standard value will also have minor effect, but a decrease in frequency of more than 5 to 10% will cause serious error and excessive heating.

Current transformers (CTs) have a secondary current that has a direct relation to the primary current. Any change in the primary current is reflected in the meters of other devices connected in series with the secondary terminals of the transformer. A window type is the most common. The conductor being monitored is placed through an insulated hole in the CT and acts as the primary of the unit. The conductor does not have to be centered in the hole; location does not affect the accuracy of reading. As with the PT, however, the burden imposed by the load does affect the accuracy.

CTs generally have a 5A secondary rating. Thus, a 600:5 ratio CT will have a secondary current of 5A when the primary current is 600A. For the operation of an ammeter, the primary current is multiplied by a factor or 1.5 and the ratio that is closest to the result is selected. For instance: if the flc= 410A; then 410 x 1.5= 615; and a 600:5A CT is used.

CTs and PTs are often located within power distribution equipment for feeding instruments and meters.

The purpose of the multiplier is to place the needle of the ammeter near the $^2/_3$ position on the scale when the current in the primary is 600A. This allows room on the scale for overload conditions.

VOLTAGE REGULATORS

Another type of specialized transformer is the voltage regulator, which maintains a constant output voltage to computers and other sensitive electronic equipment despite a wide variation in input voltage. They are devices that accepts an input from the AC power line and maintains a voltage output that is relatively constant regardless of input variations that may occur within a specified voltage range. This range is typically ± 20% nominal voltage. As these devices are relatively slow in responding, they usually do not protect against sags or swells.

There are many variations in the equipment that generally matches the generic term "voltage regulator." Each type has advantages and disadvantages

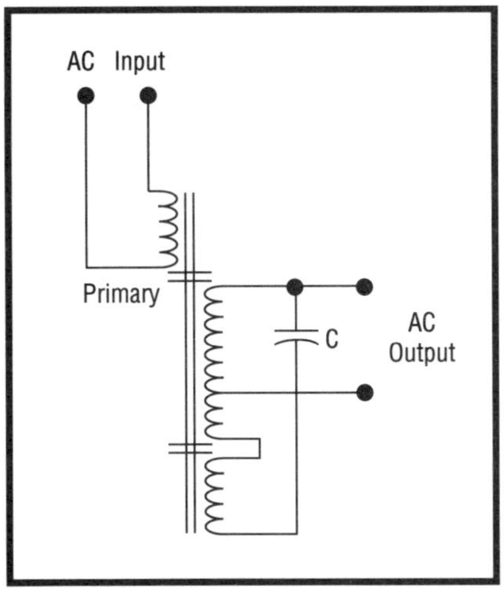

Fig. 2.12. A typical ferroresonant transformer power line regulator.

in application.

Ferroresonant transformer power line regulator can have separate input and output windings and can provide voltage transformation and common-mode noise isolation as well as voltage regulation. They provide excellent regulation, but have limited overload capacity and poor efficiency at low loads. One type (**Fig. 2.12**) has an additional pair of magnetic shunts and a filtering winding. The segment of the core under the secondary winding will saturate and ferroresonate with the AC capacitor once every half cycle, limiting the output voltage to a fixed value. The units are tuned to achieve ferroresonant regulation to the output over a broad range of input voltage. The second pair of magnetic shunts and filter winding soften the secondary core saturation effect. The filter winding is connected in series with the AC capacitor to form an LC trap to filter out the low-order harmonics generated by the ferroresonant action.

Magnetically-controlled voltage regulators are a special autotransformer type, with buck-boost windings to control the voltage. The effect of the buck-boost windings is varied by a control winding with DC current that affects the saturation of the core. The control-winding current is produced by electronic sensing and control circuits. Response time is a relatively slow 3 to 10 cycles, but response is smooth.

Electronic tap-switching transformers have thyristor switching of a tapped coil winding that changes the turns ratio. Switching is controlled by electronic sensing circuits and can be very fast (in the order of one to three cycles). Thyristor switching at voltage zero is easier and less costly than at current zero, but can cause transient voltages in the system. Current-zero switching is preferred. Voltage change is in steps, not smooth, but the steps can be small. These regulators can have the high efficiency, low impedance, noise isolation, and overload capacity of a transformer.

Data processing equipment is best served by static type regulators because of a faster response time, 10 to 20 cycles, as opposed to electromechanical regulators with a one to three second response time. In selecting the type of static regulator, consideration should be given to efficiency (this varies from 60 to 96%), total harmonic distortion (6 to 20%), accuracy (1 to 6%), power rating, cost, and interaction with the system.

The term "transformer connections" that is the subject of Chapter 3 applies to more than how feeder and load cables connect to the terminals of a transformer. The subject principally deals with the way the transformer windings are interconnected to form delta, wye, or other systems.

TRANSFORMER CONNECTIONS

3

Single-phase transformers are relatively simple to apply provided their voltage ratings are correct and they are sized adequately to carry the load. The same is not true for 3-phase transformers (or single-phase transformers connected to carry a 3-phase load). Here, the transformer must be connected into the primary and secondary distribution systems in a manner that will satisfy the system as well as the load requirements. There are many ways in which these transformers can be connected.

This chapter will deal with the fundamentals that should be known about transformer connections. Understanding terms like polarity, phasors, and phase rotation is important because they represent the basis by which transformers can the connected together as part of a power distribution system.

POLARITY, SINGLE-PHASE

When dealing with single-phase transformers, the term "polarity" is often used. A typical unit will have terminals marked H1-H2 and X1-X2, as shown in **Fig. 3.1(A)**, next page. In order to be consistent, it is necessary to standardize on a direction from which to face the terminals of the transformer. For the purpose of this discussion, in **(B)** the direction is shown by the arrow. Further, it is then required to decide the order in which *both* the high- and the low-voltage terminals are to be taken. In **(C)**, facing the terminals in the direction chosen previously, the direction is from right to left (or it could be left to right if desired).

Polarity is defined as the relative direction of induced voltage from H1 to H2, as compared to X1 to X2, both being taken in the same order (in this example, from right to left). When the induced voltages of the high- and low-voltage sides are in the opposite direction, as shown in **(D)**, the polarity is considered to be *additive*. When the induced voltages are in the same direction, as shown in **(E)**, the polarity is considered to be *subtractive*. It's important to remember that both sets of terminals must be viewed from the same side, which in this example was determined in (B).

What physically determines the polarity is the direction the conductors are wound around the core. In **Fig. 3.2(A)**, next page, is subtractive, and **(B)** is additive polarity. Transformer windings, however, are not readily available for inspection, so a test method is required. In **(C)**, two adjacent terminals, H2 and X2 (could also be H1 and X1) are connected together in series and an AC source is connected to one of the open terminals. When a voltage reading is taken across the high-voltage windings and then across the two open terminals, the reading between the open terminals will be *less* than that across the high-voltage terminals. This indicates a *subtractive* polarity. If the voltage reading is *more*, then it is *additive*. In general, it is more common for transformers to have subtractive polarities because of the lessened voltage levels between adjacent high- and low-voltage terminals.

PARALLELING SINGLE-PHASE TRANSFORMERS

Single-phase transformers are suitable for parallel operation only when their voltage ratings are identical, their percentage impedances are equal, the ratio of reactance to resistance is the same, and they have the same polarity. When these conditions are not met, there will either be an uneconomical division of current, or a circulating current will flow between the transformers. Both of these will lower the efficiency and decrease the maximum load that the transformer can safety carry. Also, if the ratios of the windings of the transformers are not equal, a constant circulating current will flow between the transformers, even at no load.

It is not good practice to operate transformer in parallel when:

- the division of load is such that, with a total load equal to the combined kVA rating, the load current flowing in any one of them is greater than 110% of its normal full-load;
- the no-load circulating current in any transformer exceeds 10% of its full-load rated value; or
- the arithmetical sum of the circulating and load currents is greater than 110% of the normal full-load current.

One of the possible sources of circulating current is that one transformer is connected to a different tap than the others. Even when only differing by one $2^1/_2$% tap, the circulating current will be significant. Assuming all have equal kVA ratings and that each have an impedance of 5%, the circulating current will be approximately 25% of the full-load current, and is independent of load and load division.

Other causes of circulating current in paralleled single-phase transformer include: impedances different in magnitude and phase angle; impedances are different and the resistances do not have the same ratio to each other as the reactances; and, where distances between paralleled units are long and the current high, the reactances of the conductors between the transformers can be significant.

POLARITY, 3-PHASE

The rules governing the connections and markings of 3-phase transformers are covered in IEEE Standard C57. Polarity, as applied to these units, indicates the sequence in which the individual windings reach their maximum positive voltage values. To help identify this polarity, the standard has adopted the numbering convention

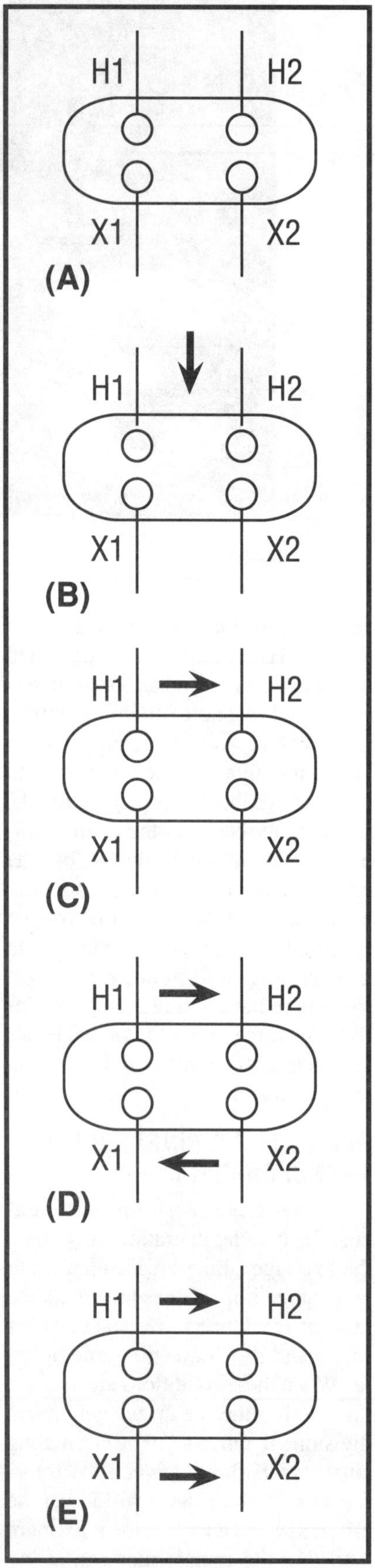

Fig. 3.1. Polarity of a transformer **(A)**, is determined by choosing a direction to view the terminals. In **(B)**, it was decided to view the terminals from the higher voltage side. The order in which to view the terminals was determined, in **(C)**, as from right to left. If the induced voltages on the lower voltage terminals are in the opposite direction **(D)**, the transformer has an additive polarity. If, on the other hand, the induced voltage is in the same direction **(E)**, the transformer has a subtractive polarity.

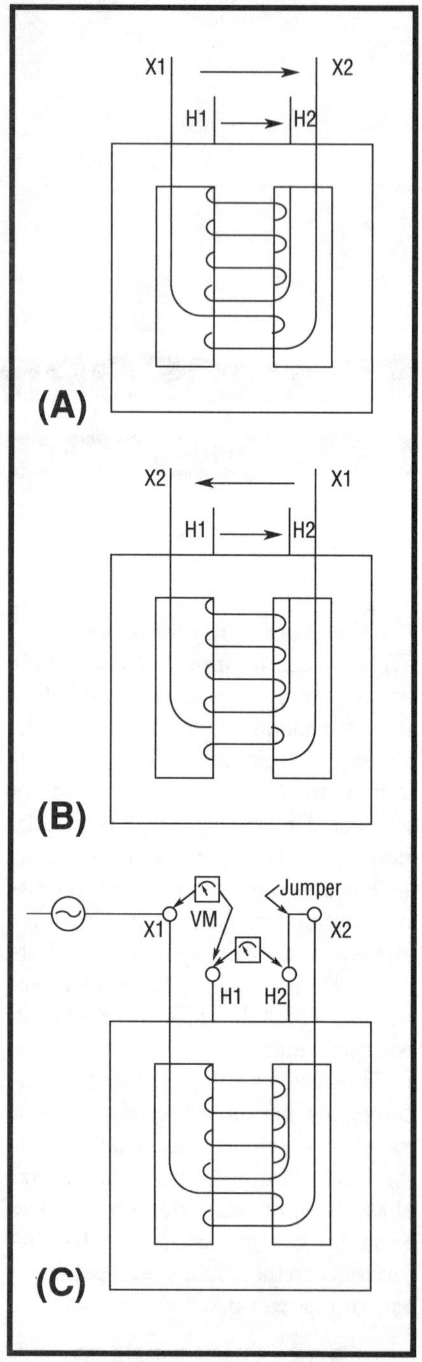

Fig. 3.2. Looking at the internals of a transformer, it is the direction in which the conductors are wound relative to each other that determines the polarity. In **(A)**, they are bound in the same direction and are subtractive; in **(B)**, they are wound in opposite directions and are additive. A method of determining polarity by an external test is shown in **(C)**.

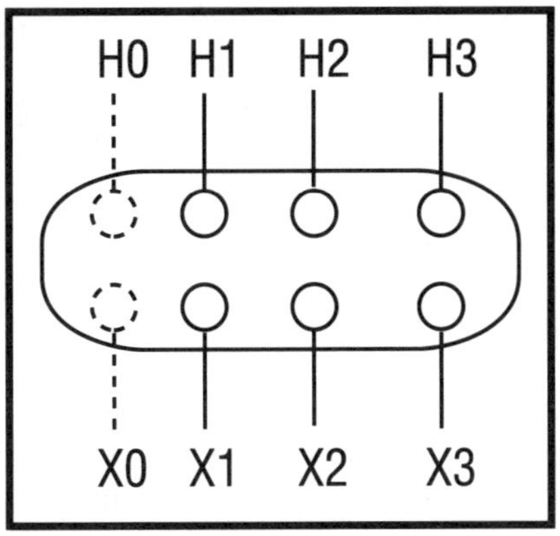

Fig. 3.3. Standard configuration on the terminals of a 3-phase transformer.

shown in **Fig. 3.3**. The terminals to which the primary winding leads are brought out are numbered sequentially as H1, H2, and H3 (H0 is the neutral, if any). Similarly, the secondary winding lead terminals are numbered sequentially as X1, X2, and X3 (with X0 being the neutral, if any).

To avoid confusion, it is also required that when H1, H2, and H3 each reach their maximum value, the corresponding low-side terminal (X1, X2, and X3) also reaches its maximum value in that sequence. In addition, as described previously for single-phase transformers, the location of the terminals shown in Fig. 3.3 indicates a subtractive polarity.

CONSTRUCTING 3-PHASE VOLTAGE DIAGRAMS

There is some confusion in terminology when analyzing how to connect 3-phase transformers in parallel. One of the essentials is developing a phasor diagram, which is often referred as a "vector" diagram. Vector is not the proper term because vectors describe quantities in three-dimensional space. Phasors, on the other hand, deal with sinewave quantities of the same frequency in electrical circuits.

Phasors are quantities that have magnitude and angular relations to represent sinusoidally varying quantities, such as AC voltages and currents, as shown in **Fig. 3.4**. Their magnitude corresponds to the voltage or current measured, and their angles represent the time differences between similar parts of different waves.

In actual practice, construction of voltage diagrams to analyze 3-phase transformers employ a standardized system. For example, in **Fig. 3.5** assume that the transformer in **(A)** is connected in delta on the high side and in wye on the low side (which is one of the more common transformer connections). A voltage phasor diagram **(B)** is drawn of the delta primary with its terminals H1, H2, and H3 in the positions dictated by the standard.

There are definite phase relationships between the connections of the primary and secondary windings. Because, per the standard, the sequences of both windings are to be the same, the voltage diagram must reflect this. Drawing the voltage diagram for the secondary windings in relation to the secondary is also standard, but it can be constructed logically.

In **(C)**, following the sequence of the delta primary, the X1 to N line for the wye secondary is drawn parallel to H1-H3, X2-N is drawn parallel to H1-H2, and X3-N is drawn parallel to H2-H3. Similar diagrams can be constructed for all 3-phase transformer connections.

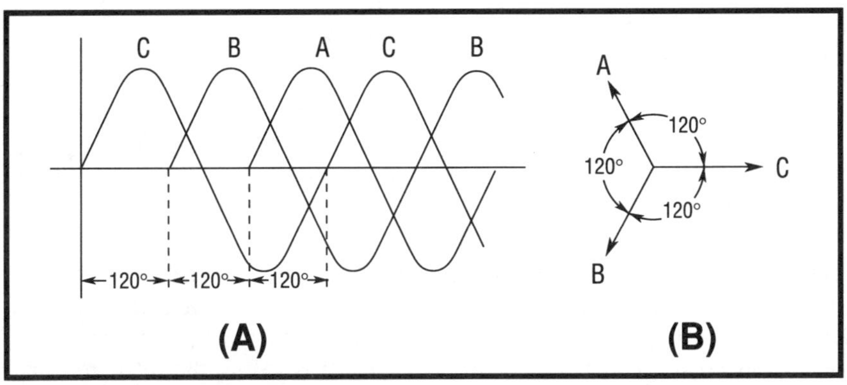

Fig. 3.4. Phasors can represent the AC voltages of a 3-phase transformer.

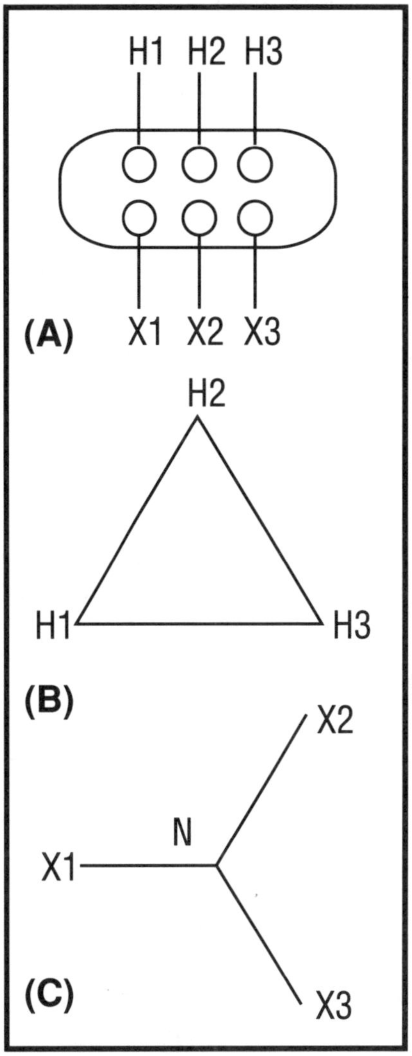

Fig. 3.5. Construction of a voltage phasor diagram of a 3-phase transformer **(A)** that is connected delta-wye. By standard convention, H1 appears on the left hand bottom, followed in sequence by H2 and H3 in a clockwise direction **(B)**. The lower voltage terminals in **(C)** appear with X1 at the left and parallel to phasor H1-H3, followed by X2 parallel to H1-H2, and X3 parallel to H2-H3.

ANGULAR DISPLACEMENT AND PHASE ROTATION

Voltage diagrams allow the angular displacement of the connections to be visualized. The angular displacement between the high-side and low-side windings is defined as the angle of each of the voltage phasor diagrams between the line passing from its neutral point through H1 and X1 respectively.

For the case a delta-wye connection shown in **Fig. 3.6(A)**, for which

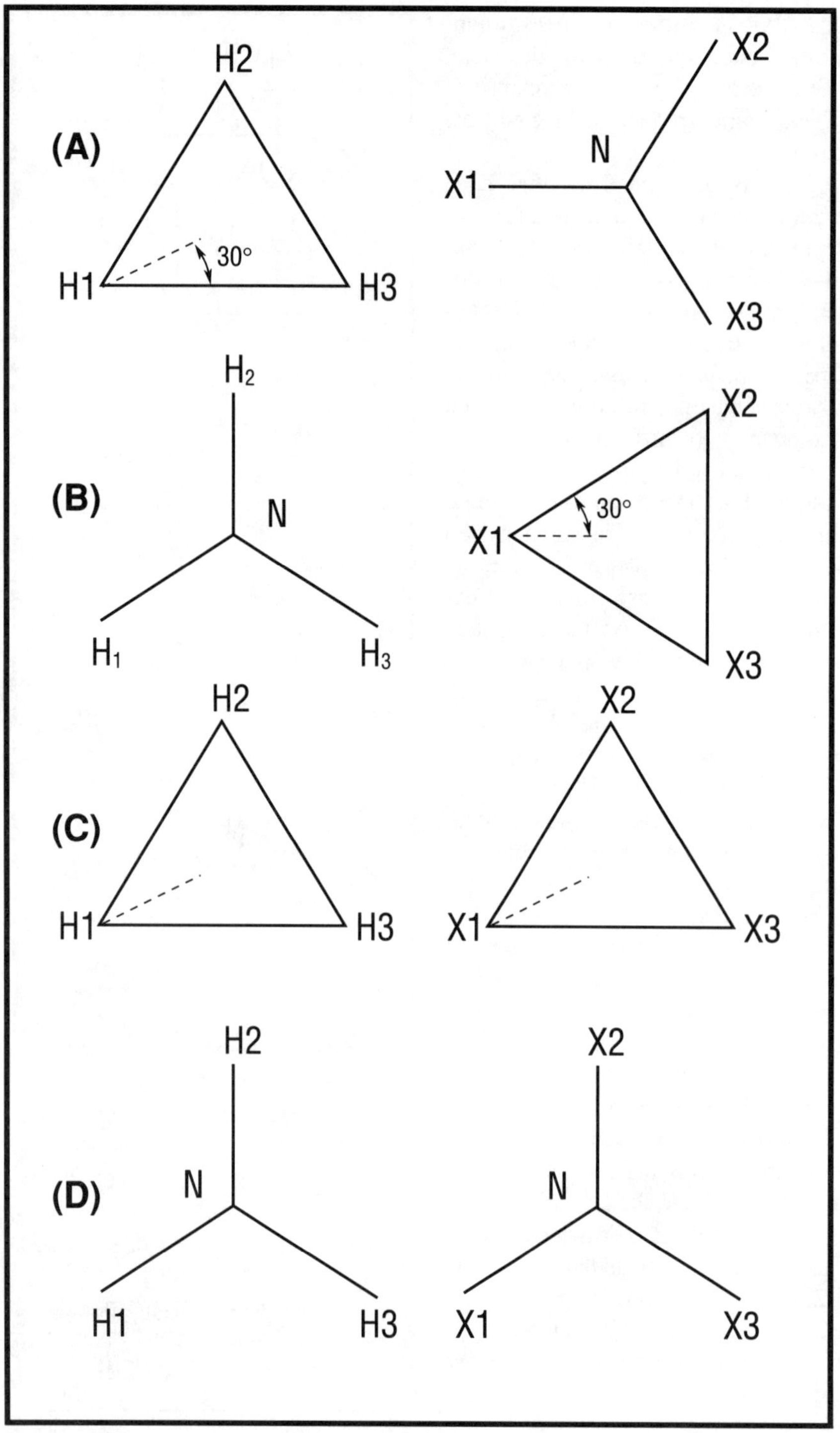

Fig. 3.6. Angular displacement is defined as the difference in the line drawn from H1-N of the high and low windings. In **(A)** there is a difference of 30° between the delta and wye connections. Similarly, in **(B)** there is also a 30° difference in the Wye-delta connections, while in **(C)** and **(D)**, no difference exists in the delta-delta and wye-wye connections.

the voltage diagram was developed previously, the neutral point of the delta connection is along a line bisecting the angle H2-H1-H3. The line X1-N is a horizontal line. The difference between the two is 30°, meaning that there is a 30° angular displacement between the high and low windings.

In **(B)** it can be seen that a 30° angular displacement also exists for a wye-delta connection. In the delta-delta connection **(C)** and wye-wye connection **(D)**, it can be seen that there is no angular displacement between the primary and secondary windings.

Due to the standardized numbering of the 3-phase terminals (H1-H2-H3 and X1-X2-X3) and the requirement that the order in which they reach their maximum positive value is the sequence of ascending numbers, facing the terminals from one side as shown in **Fig. 3.7(A)**, it can be seen that the phase rotation is counterclockwise. This, however, only indicates the *relative* phase rotation of the primary and secondary. The order in which the voltage actually appears at the terminals depends upon the rotation of the *supply* to the transformer.

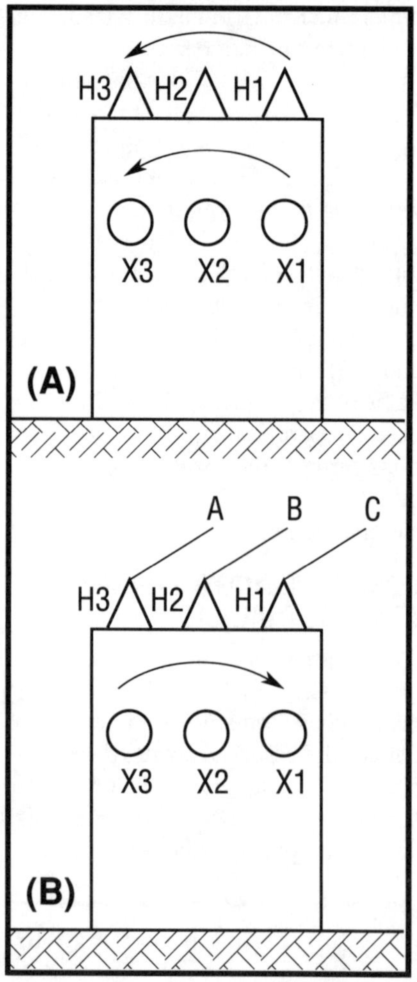

Fig. 3.7. Phase rotation is often misunderstood. In **(A)** the transformer phase rotation for this subtractive connection is counterclockwise. Actual phase rotation of the transformer connected into a distribution system **(B)** depends upon the sequence of the voltage of the line that energizes the transformer. If the sequence is A-B-C, the rotation at the output of the transformer is clockwise.

For instance, in **(B)** if the incoming line sequence was A-B-C, as connected the phase rotation at the output of the transformer would be clockwise. It would be necessary to switch two of the line leads in order to achieve a counterclockwise rotation.

DELTA AND WYE TRANSFORMER CONNECTIONS

Single-phase transformers can be connected together in delta or wye, or in any other of the 3-phase configurations. On the other hand, 3-phase transformers often are paralleled for capacity or redundancy purposes. In either case, the wye or delta connection is balanced; the currents and voltages in each of the phases are similar and related to the line voltages and currents in an identical manner. The kVA rating of the transformer with either of these two connections is equal to the kVA that can be delivered to the load (except for losses due to magnetizing current, etc.). With many of the other possible transformer connections, such as zig-zag, open-delta, Tee, and so on, the transformer will only be capable of delivering 86.6% of their rating.

Delta and zig-zag connections are the only ones with which voltage and current symmetry from line-to-line and line-to-neutral can be achieved. Wye and other connections have some degree of asymmetry that under some circumstances can lead to overvoltage conditions. Thus, Y-Y connections are not often used. The presence of the delta connection in a delta-wye connection and a wye-delta connection, for all practical cases, avoids these conditions.

An additional benefit of the delta-wye, delta-delta, and wye-delta connection is that they do not introduce third-harmonics and their multiples into the line, assuming they are carrying approximately balanced loads.

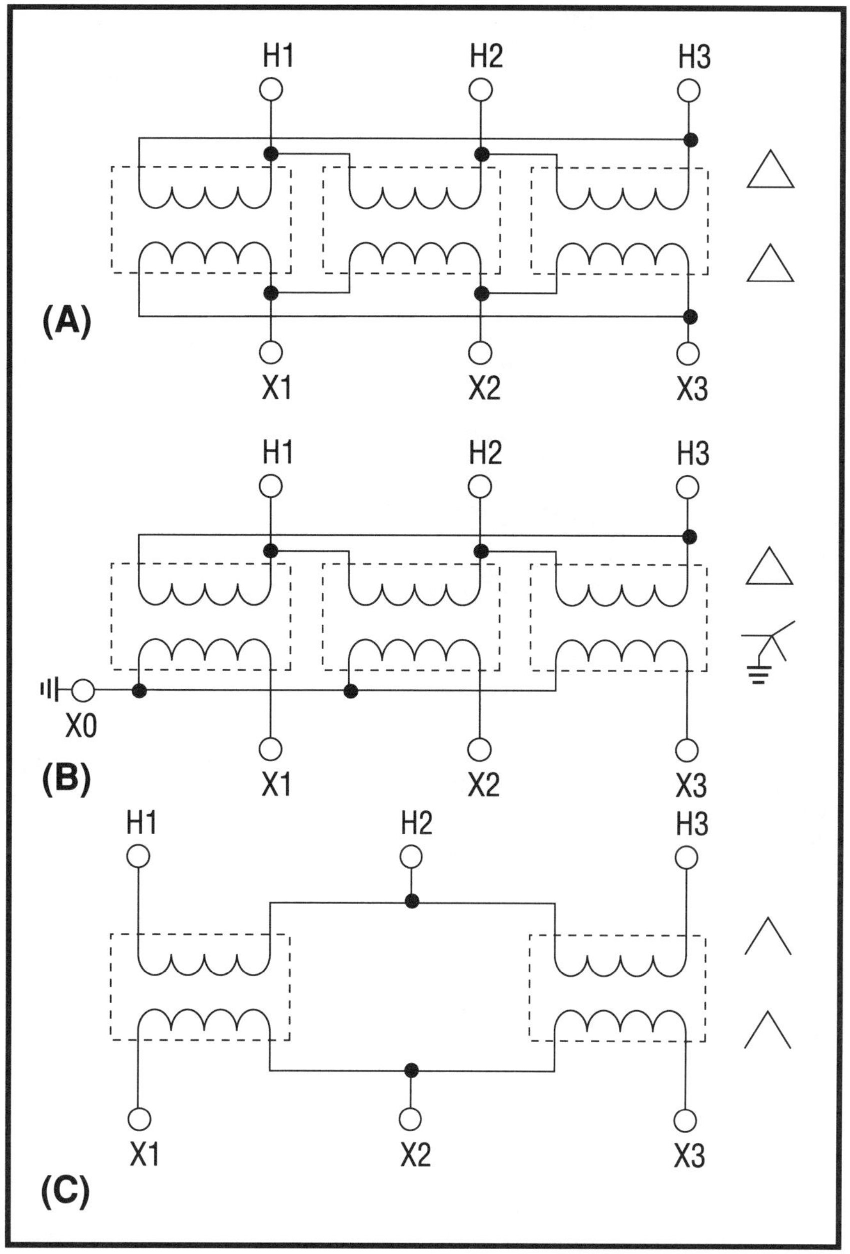

Fig. 3.8. Single-phase transformer can, after carefully matching their characteristics, be connected together to achieve a 3-phase system. In **(A)** they have been connected delta-delta; in **(B)**, delta-grounded wye; and in **(C)**, open-delta.

SINGLE-PHASE CONNECTIONS

Delta-delta, delta-wye, or any one of the other connections, can be achieved by properly tying together two or more single-phase transformers. In **Fig. 3.8(A)**, three single-phase transformers have been connected together in delta-delta; in **(B)**, they have been connected in delta-wye; and in **(C)**, two single-phase transformers have been connected in open-delta to serve a 3-phase load.

An open-delta connection is sometimes used to provide power to 3-phase loads when one of the transformers of a bank made up from three single-phase transformers has failed. It also is employed as a temporary replacement when a 3-phase transformer has failed and only two single-phase transformers are immediately available. What is interesting in this case is that the impedance of the transformers do not have to be the same while operating in this configuration. If, however, a third transformer is to be added later to achieve a closed-delta connection, the impedances *must* be equal. It also should be noted, that transformers in an open-delta configuration can only deliver 86.6% of the kVA rating of the sum of the rated kVA of each unit.

When connecting single-phase units a delta-delta configuration, the voltages are determined by the circuit, but the division of current among the phases is determined by the internal characteristics of the transformers. In a delta-wye configuration, the division of current is the same regardless of the differences in individual transformers. For example, a balanced 3-phase load will be equally divided among the phases of delta-wye banks regardless of differences in impedances in the transformers, while with a delta-delta bank the load will be equally divided only if all impedances are equal.

GROUNDED VS UNGROUNDED SYSTEMS

One of the principal benefits of using a delta-wye connection is that a neutral is created that can be grounded. When a ground fault occurs, the phase-to-ground voltage of the faulted phase does not fall to zero, and the phase-to-ground voltages of the other two phases are not greatly affected by the fault. The current in a faulted phase conductor always flows to ground through the fault and returns from ground via the grounding electrode conductor at the X0 transformer terminal. All faults are downstream from the system ground point so all ground-fault sensors upstream from the fault will detect the fault, but those downstream will not. This makes locating the fault easier. Another obvious benefit of a wye secondary is that line-to-neutral loads can be carried operating at a voltage equal to the line voltage divided by the square root of 3.

Ungrounded systems, such as a delta-delta connection, have no provision for grounding and, thus, the phase-to-ground voltages are not fixed. As a result, when a ground fault occurs, the faulted phase assumes approximately ground potential and the fault itself becomes the system ground point. Because there is no other conductive path to ground, there is no heavy flow of current through the fault. Thus, the system can continue to operate under these conditions. Ground-fault indicators can be employed to warn of the presence of a grounded phase. This will allow location and correction of the fault without interruption to the remainder of the system. The fault, however, must be cleared before another phase experiences a ground fault, which creates essentially a phase-to-phase fault that will trip the overcurrent protective device.

There are several ways in which delta-delta systems can be protected from ground faults, including the application of a grounding transformer to create and artificial neutral. Relaying is then applied to this neutral to cause the overcurrent protective device to open.

GROUNDING AUTOTRANSFORMERS

Zig-zag autotransformers are installed in 3-phase, 3-wire ungrounded systems to create a 3-phase, 4-wire distribution system, or to provide a neutral reference for grounding purposes.

Where a grounding autotransformer is connected to a 3-phase, 3-wire sys-

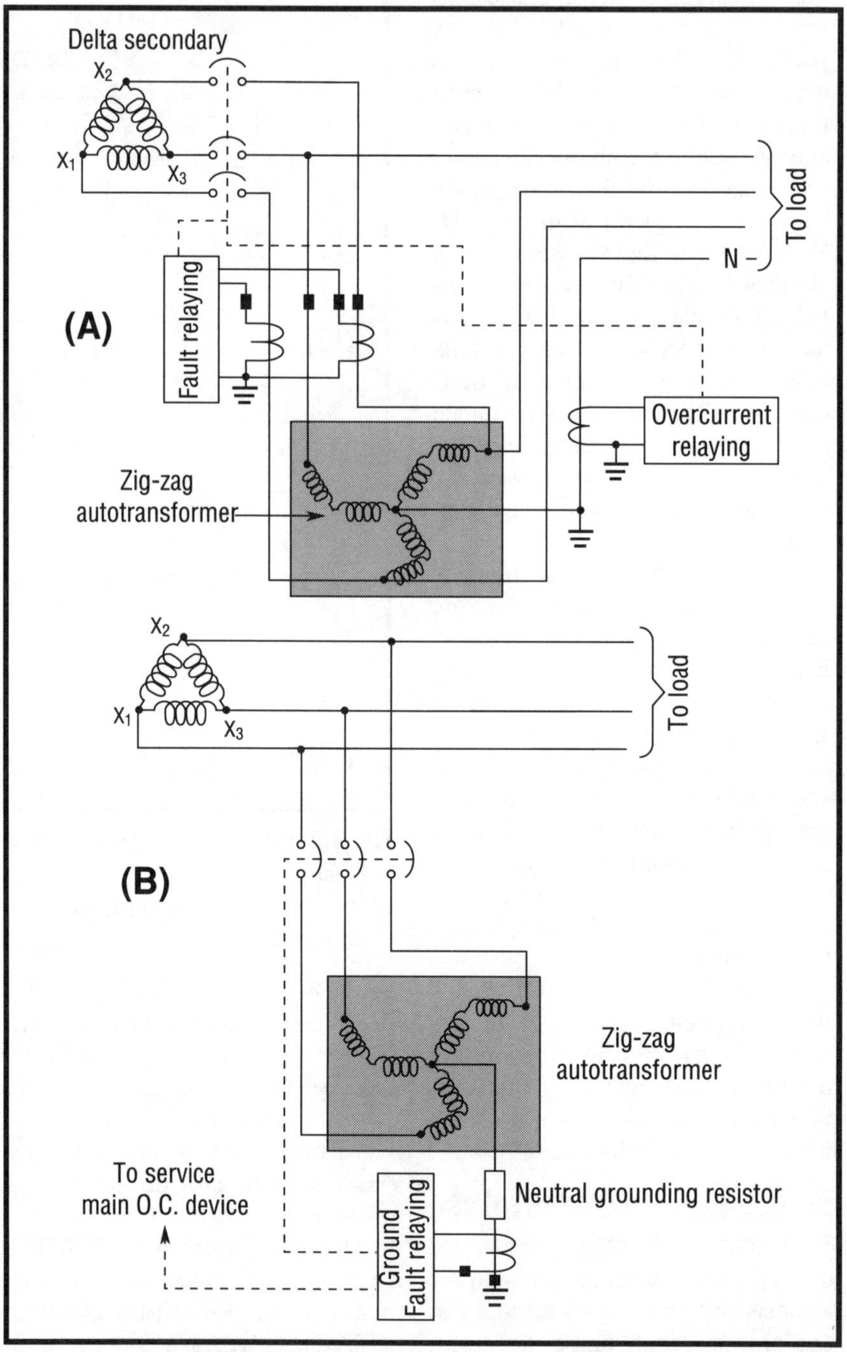

Fig. 3.9. Zig-zag grounding autotransformers used to derive a neutral from a delta-connected secondary of a transformer **(A)**, and for ground-fault detection in **(B)**.

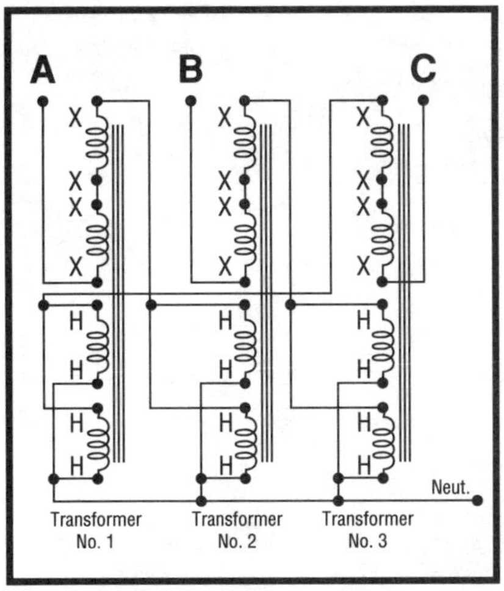

Fig. 3.10. Connections required to achieve a zig-zag configuration.

tem to create a system from which phase-to-neutral loads can be supplied, as shown in **Fig. 3.9(A)**. Relaying must be provided for the autotransformer to cause the opening of the main switch or overcurrent device for the 3-phase, 4-wire system if single-phasing or an internal fault occurs.

Ungrounded 3-phase, 3-wire systems can also use a zig-zag grounding autotransformer to provide protective relaying against ground-fault current. In these cases, a grounding autotransformer must be used to derive a neutral as shown in **(B)**.

A zig-zag grounding autotransformer is made up of three standard single-phase transformers connected as shown in **Fig. 3.10.** With balanced 3-phase loads, there is no current flowing through the autotransformer. When the loads are unbalanced, the unbalance current in the neutral wire divides equally, with $^1/_3$ flowing in the three phases. Because of its characteristics, the rating of zig-zag grounding autotransformers is based on the line voltage and the unbalanced current in the neutral, not on the balanced 3-phase load.

T-CONNECTION

The T connection (**Fig. 3.11**) uses two transformers to supply power from a 3-phase source to a 2-phase, 3-wire load. The "main" transformer is connected line-to-line at terminals H1 and H3; the "teaser" transformer is connected from the midpoint of the main to what would normally be the H1 terminal of the teaser. The H3 terminal of the teaser is connected to the third phase. The secondary could be used to carry a 3-phase load, or single-phase loads from the secondary lines to the neutral X0.

Like the open-delta connection, the T connection can be used in emergency situations where one single-phase transformer of a 3-phase bank has failed, the remaining ones can be connected in T. However, the centerpoint of both the primary and secondary windings of the main transformer must be available to make the connections.

If the transformers used to make up the T connection are duplicates (such as when they were part of a 3-phase delta connection), the new connection will carry 15.5% *greater* current than the standard single-phase transformer of the same kVA and voltage. This can be overcome by using the 86.6% tap on the teaser transformer.

If the teaser transformer is not a duplicate of the other transformer, then it can have a rating of one-half of the kVA required.

OTHER CONNECTIONS

There are many connections other than the ones described in this chapter. The versatility of transformers allow hookups to solve almost all conditions met in the typical power distribution system. The ones described here, however, are the ones most commonly found.

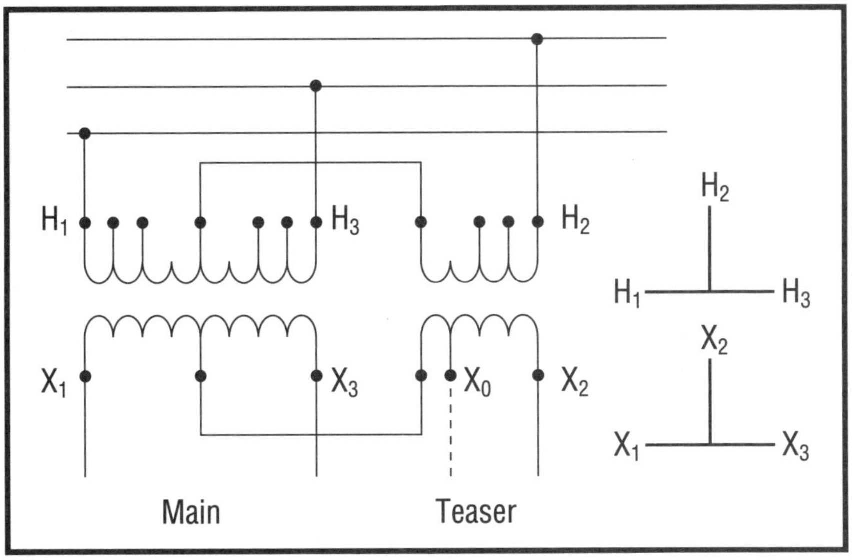

Fig. 3.11. The T connection using a main and a teaser transformers.

Power distribution in an industrial manufacturing facility employs virtually all types of transformers, from the largest power transfomers to instrumentation CTs and PTs.

TRANSFORMERS IN POWER DISTRIBUTION SYSTEMS

4

Transformers are not stand-alone items. They must be integrated with other electrical components to form a distribution system. Because of their specialized nature, there are many calculations required and rules of the National Electrical Code that must be adhered to in order to assure the safety and reliability of the circuits involved. In this chapter, some of the typical design considerations will be presented.

CONNECTION IN A SIMPLE RADIAL SYSTEM

The basic step-down transformer is usually applied in a power distribution system to reduce the voltage from that supplied by the serving utility to the level used to distribute power within a facility. Additional transformers are provided to reduce the distribution voltage level to some lower voltage required by the specific loads.

For example, in the simple radial system shown in **Fig. 4.1**, assume the incoming power source is 13.8kV, 3-phase, from the utility company's overhead poleline. Transformer 1 is used to reduce this voltage from 13.8kV delta to 480V delta for distribution within the facility via the switchgear. Feeder No.1 goes to a motor control

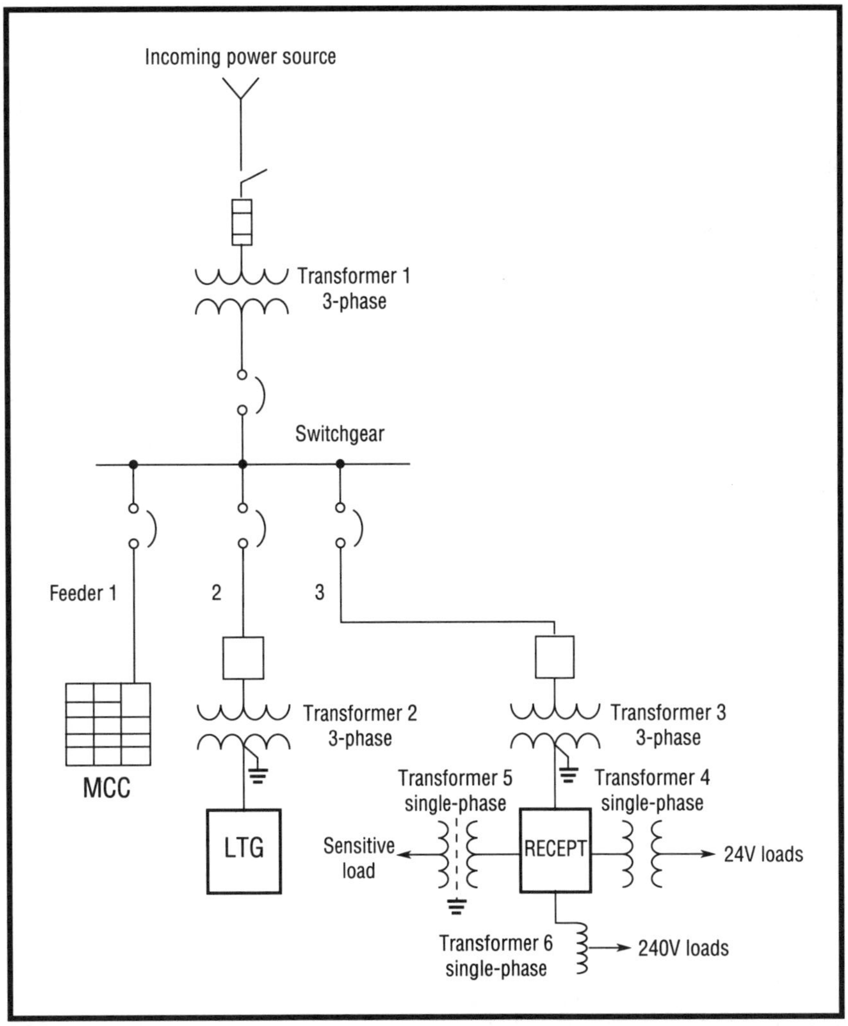

Fig. 4.1. Application of transformers within a radial power distribution system.

center (MCC) where this utilization voltage level is used directly for 3-phase 480V motor loads.

Feeder No. 2 serves Transformer 2, which has a 480/277V wye secondary (three phases plus a neutral) to supply power, via the panelboard, to 277V fluorescent lighting. The main purpose of this transformer is to derive the neutral, rather than to reduce the voltage.

Feeder No. 3 serves Transformer 3, which lowers the voltage from 480V to 120/208V to feed receptacles and other loads. Here, the transformer is used both to lower the operating voltage, and also to derive a neutral.

Among the loads fed from the receptacle panelboard is one that requires a single-phase 24V feed, which is provided by Transformer 4. Another single-phase load involves sensitive electronic equipment, which includes personal computers, printers, etc. Here the need is to prevent transient voltages from upsetting their operation. A shielded isolation transformer (Transformer 5) is installed for that purpose. A 240V load must also be fed from this panelboard. A buck-boost autotransformer (Transformer 6) has the capability of raising the voltage from the available 208V to the needed 240V utilization voltage.

Integrating the transformers into the system is relatively simple. The first order of business is to determine the voltage, phase, and frequency requirements of the loads and grouping together those loads with the same needs. The total kVA requirements for each of these groups are then calculated by adding together the requirements of the individual load. The kVA rating of the transformers are then specified to accommodate the load, with provisions to accommodate future load growth.

Transformers come in a very broad range of voltages, ratings, insulations, and specialties. Specification of transformers is covered in more detail in Chapters 9 and 10 of this book. Because overcurrent protection is such an important item and the NEC rules that apply to transformers are so involved, this subject is discussed at length in Chapter 5 of this book.

PARALLEL OPERATION

Transformers are not restricted to being applied singly. They are permitted to be connected in parallel and switched as a unit provided that each one is protected by the proper overcurrent device. Proper, in this case, means that they are protected according to the requirements of NEC Sec. 450-3(a)(1) for transformers connected to a system operating at over 600V, or Sec. 450-3(b)(2) for those connected to systems operating at lower voltages. Both of these requirements are covered in Chapter 5.

There are several reasons why transformers are paralleled. One is that the load has grown to the point that it exceeds the rating of an existing transformer. Another is to guarantee service continuity in the event one of the transformers fails.

What must first be understood is the meaning of the word paralleling when applied to transformers. When two or more transformers feeding a common load are supplied from different supply circuits, this *is not* parallel operation. The reason is that system impedances other than transformer impedances influence load division. When two or more transformers have one set of terminals connected to a common supply bus and another set of terminals connected to a common load bus they are, by definition, paralleled.

In order for transformers to be paralleled successfully, they must have approximately equal turn ratios and impedances. At zero load, if the transformer ratios do not match, a circulating current must flow to compensate for the difference in induced voltages at the transformer terminals (**Fig. 4.2**). The resulting IZ drop adds to the lower voltage and subtracts from the higher voltage. The magnitude of the circulating current is directly proportional to the ratio difference and inversely proportional to the sum of the transformer impedances.

If paralleled transformers having unequal ratios are loaded, a current circulates. If the paralleled transformers are loaded and the turn ratios are equal, there is no circulating current and the transformer impedances alone determine the load division.

Load division is another important factor that must be considered when transformers are to be paralleled. A rule-of-thumb holds that when a load equal to the combined kVA rating of the units is applied, the load current flowing in any one unit should be less than 110% of its rating and that the no-load circulating current in any one transformer should be less than 10% of the rated current.

Of equal importance when transformers are to be paralleled is that they must be properly connected. In single-phase transformers, the instantaneous voltages must be in the same

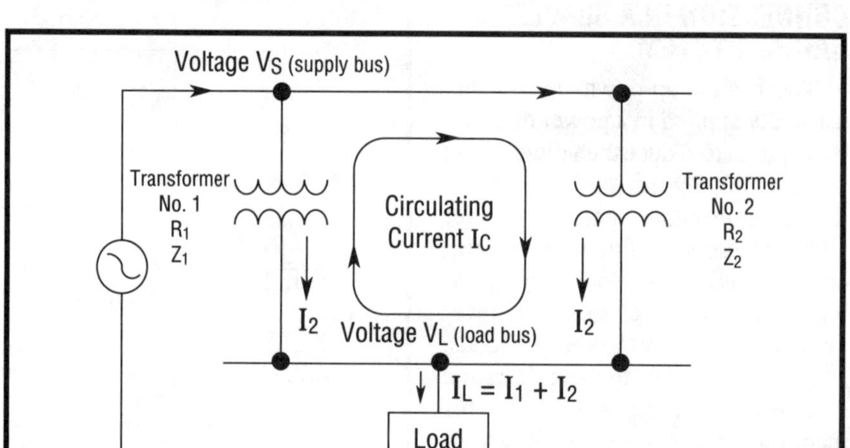

Fig. 4.2. Unless the turn ratios and impedances of paralleled transformers are equal, a circulating current (I_c) will flow. In this example, transformer 2 has a lower turn ratio than transformer 1, setting the direction of the circulating current flow.

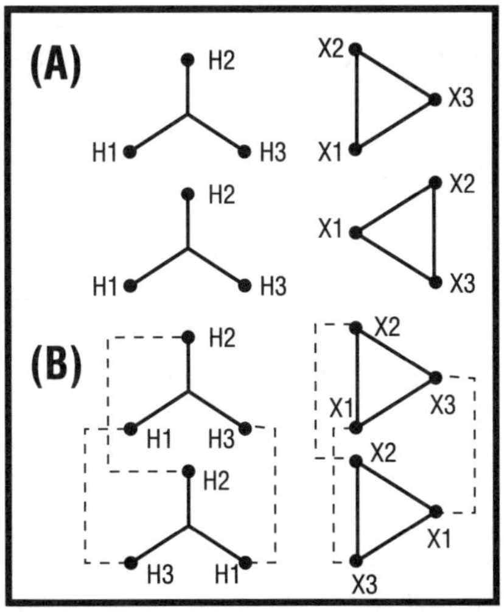

Fig. 4.3. Phasor diagram of two transformers **(A)** that are to be paralleled, and the proper connections that must be made **(B)**.

direction (H1 to H1, H2 to H2, X1 to X1, and X2 to X2). When paralleling 3-phase banks, in addition to their polarities, ratios, and impedances, the angular displacement between windings and the connection sequence must be such that the transformers have the same phase rotation. **Fig. 4.3(A)** shows a phasor diagram of two transformers that are to be paralleled, and **(B)** shows the proper connections.

TRANSFORMER NETWORKS

Grid-type secondary network systems similar to the one shown in **Fig. 4.4** have been used extensively by electric utility companies serving high-density load areas in cities. This arrangement assures the maximum reliability in powering loads. Depending upon the sizing strategies employed, upon the loss of one or more of the transformers, those remaining on the network can pick up all or part of the load. The NEC gives specific rules for these networked transformers.

When secondary networks are used, an overcurrent device rated or set at not more than 250% of the rated secondary current of the transformers must be provided in the secondary connections of each transformer. In addition, an automatic circuit breaker actuated by a reverse-current relay set to open the circuit at not more than the rated secondary current of the transformer must be provided in the secondary connection of the transformer.

In general, both requirements are usually provided by the installation of a "network protector" as shown in **Fig. 4.5**. A network protector is a complete unit that consists of a power circuit breaker, relays, and fuses, sometimes packaged with the stepdown transformer. The network protector breaker:

- closes if the supply and network voltages are such that, if closed, the protector transformer will deliver power to the network;
- trips on approximately 0.1% of rated load or fault current flow from the network to the supply; and
- will not operate for network faults, but the fault can be cleared by the protector fuses if the fault does not burn itself clear.

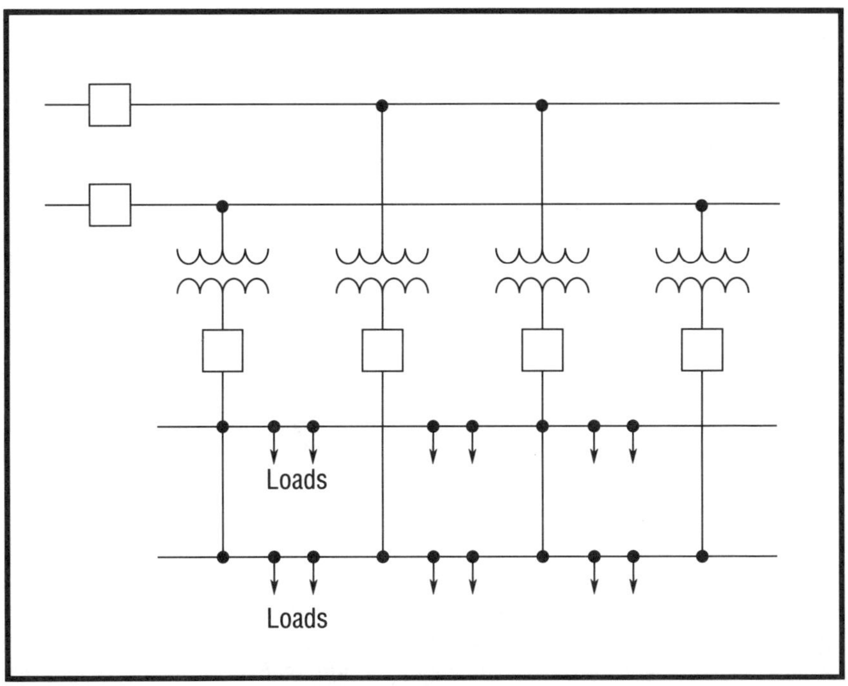

Fig. 4.4. A typical distributed (grid-type) secondary network arrangement.

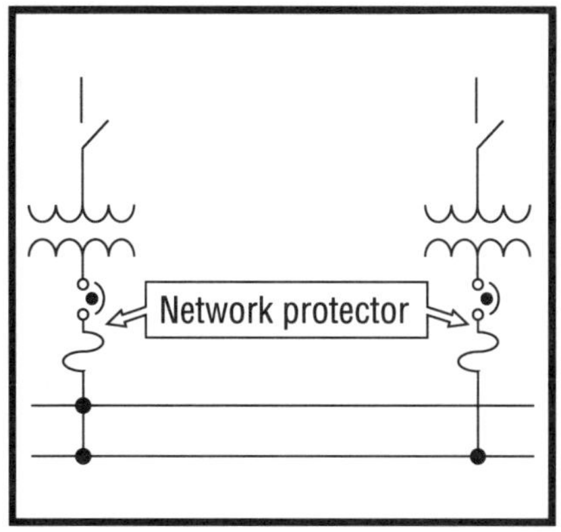

Fig. 4.5. Provisions must be made to prevent backfeeding power in a networked system. Network protectors usually are used to provide this protection.

The main purpose of this requirement is to prevent backfeed when one of the transformers is taken out of service for maintenance or due to a fault. The transformer(s) remaining on the line will not only feed the load, but will also energize the secondary of the out-of-service transformer unless it is completely isolated from the system. The network protector automatically opens the circuit when a current is sensed flowing *from* the network *to* the transformer.

SECONDARY TIES

A secondary tie is defined as a circuit operating at 600V, nominal, or less, that connects two power supply points. The connection is usually cable between the secondary of two transformers, as shown in **Fig. 4.6** (next page). The intent is that upon the failure of one unit, the tie can be closed allowing the remaining transformer to carry the entire load. Sometimes such an arrangement requires the shedding of some of the load in order to maintain the most critical loads in operation.

A secondary tie can be made up of

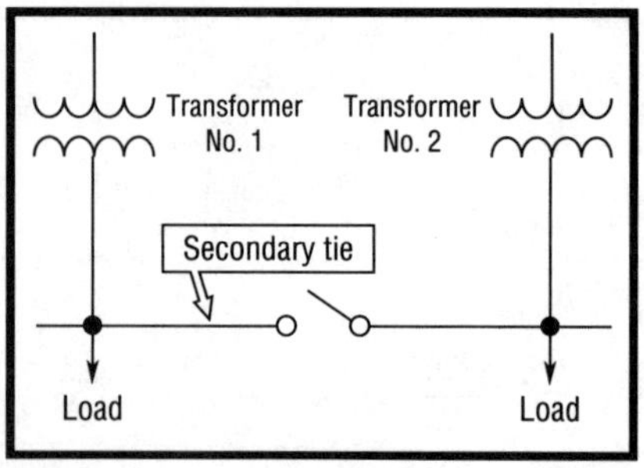

Fig. 4.6. A typical transformer secondary tie.

one or more conductors per phase. The NEC gives specific details on how secondary ties are to be arranged and protected.

The general rule for a secondary tie is shown in **Fig. 4.7**. An overcurrent protective device must be installed at *each end* of the tie circuit. The permitted protective devices are: a fuse, circuit breaker, or a combination of current transformer and overcurrent relay connected in series with each ungrounded conductor; or listed devices providing equivalent overcurrent protection in closed-loop power distribution systems.

Where the tie circuit is made up of two or more paralleled insulated cables per phase, devices called "cable limiters" are permitted to be installed at both ends of each tie conductor. They open at a predetermined temperature of the tie conductors under short-circuit conditions. Limiters are a specialized form of fuse that are not intended to protect the conductors against overload conditions, but rather to prevent their temperature from reaching a value that will damage their insulation should a short circuit occur.

There is a general NEC rule that a disconnect must be provided for fuses on their supply side in circuits operating at over 150V to ground, or where cartridge fuses of any voltage are accessible to other than qualified persons. Secondary ties provided with limiters must have a switch at each end of the tie, that when open, deenergizes the associated tie conductors and limiters.

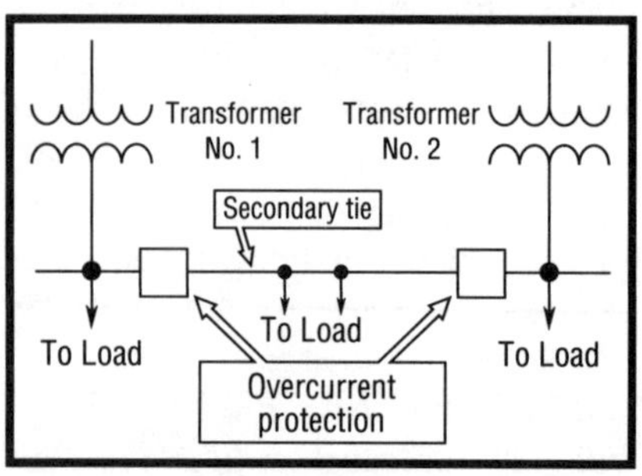

Fig. 4.7. General requirement for O.C. protection of secondary ties.

Sometimes loads can also be fed directly from a secondary tie. When loads are connected to the secondary tie, additional NEC rules must be followed very carefully. For more details on the applicable NEC rules, see the *EC&M* publication entitled, *Understanding NE Code Rules on Transformers*.

PHASE-SHIFT PROBLEMS

When powering a motor from separately transformed power sources as shown in **Fig. 4.8**, an uninterrupted transfer operation may result in serious phase shift problems if certain safeguards are not employed. Even with verification of correct motor rotation on each source, the sources can be 60° out of phase at the transfer switch. A load fed from two feeders that are to be paralleled, even momentarily, requires an exact match of system phases.

How can the two feeders at the transfer switch be 60° out of phase? Recall the explanation of voltage phasor diagrams in Chapter 3. For Feeder No. 1, the primary leads to transformer T1 are connected A-B-C to H1-H2-H3; the primary phase sequences are A-B-C to H1-H2-H3; the secondary phase sequences are X1-X2-X3; and the low-side voltages lag the high-side voltages. For Feeder No. 2, the primary leads to transformer T2 are connected C-B-A to H1-H2-H3; the primary phase sequence is still A-B-C, but, the primary terminal phase sequence is H3-H2-H1; the secondary phase sequence is X3-X2-X1; and the low-side voltages lead the high-side voltages. Phase A is connected to X3 and Phase C to X1. This produces the phase relationships at the motor as shown. Rotation will be verified as

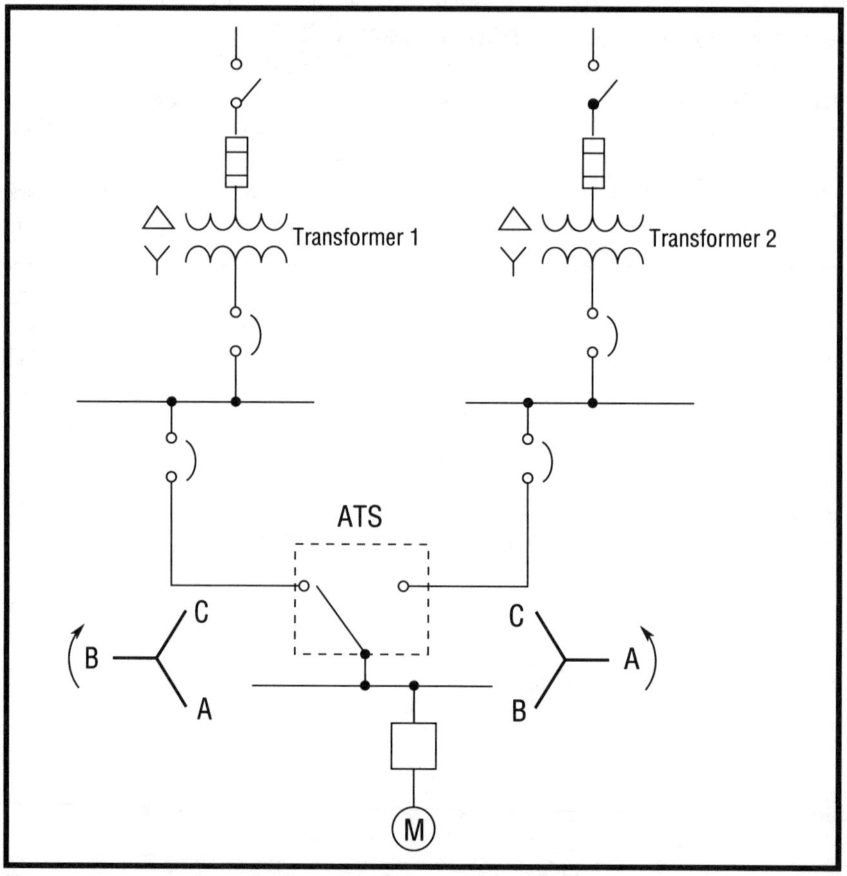

Fig. 4.8. Single-line diagram of power distribution from separate delta-wye transformers. Instantaneous phase relationships for both feeders at transfer switch are shown by phasor diagrams.

correct, but, the phases of both feeders will be 60° apart. Since the secondary rotation at T2 is backwards, the secondary cables need to be "rolled" to correct for this.

To avoid connecting two incompatible sources, a checkout procedure could discover this error. Phase rotation or sequence is defined by ANSI as "the order in which the voltages successively reach their positive maximum values." The normal method for checking correct phase rotation on a motor is to "bump" it. If it runs backwards, then reverse any two leads. But that does not mean the system is properly phased out. Functionally, for a single power source, it does not matter. If, however, the connection between two sources is to be checked for proper phasing, not only must the correct rotation be verified but the phases must be exactly matched. Correct angular displacement must be verified.

A simple method, using a volt meter, can be used to verify correct phasing between two sources. With the voltmeter connected between Phase A on Source 1 and Phase A on Source 2, voltage measurements are taken. Readings are also taken, in the same manner, between respective B and C phases on both sources. These readings should all be zero volts. If the readings are 554V, 480V, or 277V, there is a wiring error.

A method for confirming the above voltmeter operation and results is to read the voltages between Phase A, Source 1 and all three phases on Source 2. Correct readings will be: zero volts to Source 2, Phase A; 480V to Source 2, Phase B; and 480V to Source 2, Phase C.

Line up of transformers and MV service and metering switchgear are typical of those applied at large urban facilities.

OVERCURRENT PROTECTION OF TRANSFORMERS

5

All electrical equipment must be protected against the effects of high current flow during a fault in the electrical system. NEC rules that specifically apply to power transformers are of great importance and, thus, are covered in depth in this chapter. These rules cover most transformers, including those installed in classified (hazardous) locations.

Note that this chapter also includes the rules applying to overcurrent protection of transformers used to provide power to a control circuit of a motor controller (starter). These types of transformers are commonly referred to as "control-circuit transformers," even though they are more accurately call "machine tool transformers" as described in Chapter 2. The rules for control-circuit transformer are in addition to, and sometimes supersede, those for other power transformers.

The rules covered in this chapter, however, do not apply to the following types of transformers:

- potential transformers (PTs)
- current transformers (CTs);
- transformers used in Class 2 or Class 3 circuits;
- transformers for power-limited fire-protective signaling circuits;
- dry-type transformers that are a component part of a listed piece of apparatus;
- transformers that are integral parts of X-ray, high-frequency, or electrostatic-coating apparatus;
- transformers for sign and outline lighting;
- transformers for electric-discharge lighting; and
- transformers used for research, development, or testing, where effective arrangements are provided to safeguard persons working with the equipment from contacting energized parts.

LARGE POWER TRANSFORMERS

Rules covered here apply to a bank of single-phase transformers connected to operate as a single unit, as well as to individual single-phase or 3-phase units operating alone.

Where an overcurrent device on the transformer *secondary* is required by these rules, it can consist of not more than six circuit breakers or sets of fuses grouped in one location. Where multiple overcurrent devices are used, the total of all the device rating must not exceed the allowed value of a single overcurrent device. If both breakers and fuses are used as the overcurrent device, the total of the device ratings must not exceed that allowed for fuses.

It should be noted here that nonlinear currents can increase the heat in a transformer without operating the overcurrent devices. This is a complex subject that is covered in depth in Chapter 7 of this book.

TRANSFORMERS OVER 600V, NOMINAL

These transformers must have primary and secondary protective devices rated or set at not more than the value of transformer rated current noted in the table in **Fig. 5.1** (next page). Electronic fuses that can be set to open at a specific current are to be treated similar to circuit breakers.

Medium voltage secondary. The subject has to be divided into two pieces, because the calculations required for these transformers vary according to whether the voltage of the secondary is above or below 600V. This first example assumes a medium voltage primary and secondary.

EXAMPLE: As seen in **Fig. 5.2** (next page), a 2000kVA, 3-phase, 26.4 to 2.4kV transformer has a 5.9% impedance. What size fuses would be required to protect the primary of the transformer, and what size circuit

Transformer rated impedance	Primary		Secondary		
	Over 600 volts		Over 600 volts		600 volts or below
	Circuit breaker setting	Fuse rating	Circuit breaker setting	Fuse rating	Circuit Breaker setting or fuse rating
Not more than 6%	600%	300%	300%	250%	125%
More than 6% and not more than 10%	400%	300%	250%	225%	125%

Fig. 5.1. Maximum rating or setting for overcurrent protective devices for transformers connected to a system operating at over 600V. Derived from NEC Table 450-3(a)(1).

breakers would protect the secondary of the transformer?

ANSWER: The current is obtained either from the nameplate or the equation:

$$A = kVA \div (V \times 1.73).$$

In this example, the primary current is approximately 44A, and the secondary current is approximately 482A. Based on these currents, the maximum rating of the primary fuses for a system operating at over 600V, per Fig. 5.1, is:

$$44 \times 3 = 132A$$

and the maximum setting of the circuit breaker protecting the secondary with an operating voltage over 600A is:

$$482 \times 3 = 1446A.$$

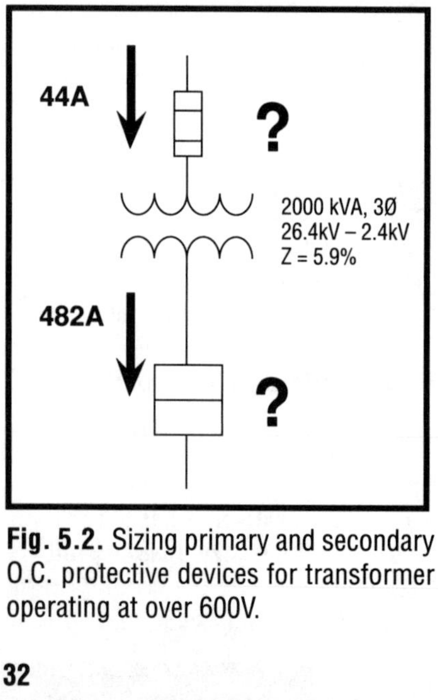

Fig. 5.2. Sizing primary and secondary O.C. protective devices for transformer operating at over 600V.

A general rule says that if a required fuse rating or circuit breaker setting does not correspond to a standard rating or setting, the next higher standard rating or setting is permitted. In this example, 150A fuses can be used for the primary because they are the next higher standard ampere rating listed in NEC for fuses.

Determining the size of the secondary breaker is somewhat more difficult. If the NEC values for standard ratings of circuit breakers are employed, a 1600A breaker would be used. This does not comply with what is actually available. For 5kV Class circuit breakers, standard values are 1200A and 2000A. Since a 1200A CB is too low, then 2000A (which is also listed in the NEC as a standard size) would be selected as the maximum rating.

Low-voltage secondary. The preceding example assumed a medium voltage secondary operating voltage. The calculation of the maximum rating of the circuit breaker if the secondary voltage was 600V or less are somewhat different. Assuming the impedance is not more than 10%, according to the table the secondary amperes would be permitted to be multiplied by a factor of only 1.25.

If the secondary 600V breaker is nonadjustable, then the next highest rating would apply as the maximum if the calculated value did not correspond to a standard rating. If, however, the secondary circuit breaker has an adjustable long-time pickup setting (such as found on low-voltage power circuit breakers), then its maximum setting is required to be used as the rating of the circuit breaker. Standard frame sizes of these circuit breakers are generally 800, 1600, 2000, 3200, and 4000A, and the maximum trip settings usually correspond to the frame size.

Where the adjustable-trip circuit breaker is located in a room or enclosure that is locked and accessible only to qualified personnel, then the long-time pickup *setting* can be used as the rating of the breaker.

The same hold true for the case where the secondary is protected by a molded-case circuit breaker that has adjusting means protected from tampering by sealable covers. The circuit breaker ampere rating is permitted to be equal to the *adjusted* long-time pickup setting. An example of this would be a typical CB with a solid-state trip unit having a continuous current (long time) adjustment from 20 to 100% of its maximum trip unit rating. If, for example, the breaker had a 800A frame and an 800A trip unit, it would be permitted to use a trip setting of 800A (100%) as the maximum. It would also be permitted to set the trip at 720A (90%) or an even lower value, as long as unqualified persons could not change that setting.

There is the other rule on overcurrent protection of transformers that needs to be considered. Suppose the transformer secondary overcurrent protection is provided by more than one device or set of devices (up to a total of six. The total of all the device ratings must not exceed the allowed value calculated assuming the secondary is protected by a single overcurrent device. If they consist of a mix of CBs and fuses, then the total of the devices must not exceed that allowed for fuses.

EXAMPLE: As shown in **Fig. 5.3**, a 500 kVA transformer has its secondary protection provided by individual circuit breakers for three 150A feeders, along with a CB for a 75 hp motor branch circuit. Is this an acceptable arrangement?

ANSWER: The secondary current of the transformer is 600A and the

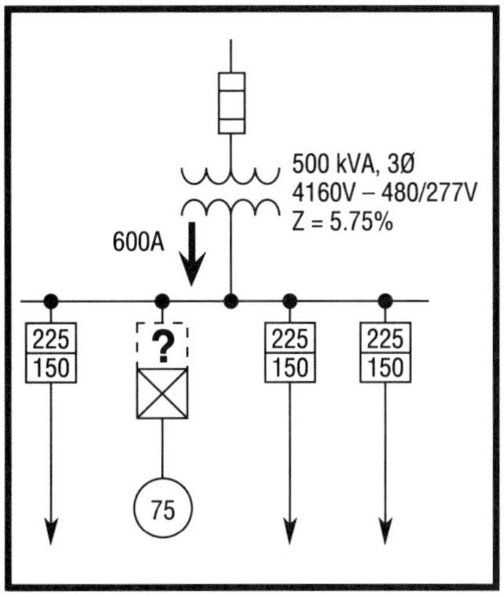

Fig. 5.3. More than one O.C. protective device on secondary of transformer operating at over 600V.

maximum permitted setting of the secondary circuit breaker is 750A.

According to the NEC, the 75 hp motor has a full-load current of 96A. Based upon the total load (150 + 150 + 150 + 96 = 546A), the configuration is adequate. To allow for the starting of the motor, however, which involves typical inrush currents of six times the motor flc, an adjustable instantaneous trip (magnetic only) circuit breaker is used. According to the code, these units may be set to trip at a maximum of 700% the motor flc, and is allowed to be raised to 1300% when the seven times setting does not allow the motor to start without tripping the device. Under almost all circumstances, it is likely that the combined total of the settings of the overcurrent devices will exceed the permitted 750A for a single secondary overcurrent protective CB, and would not be according to code.

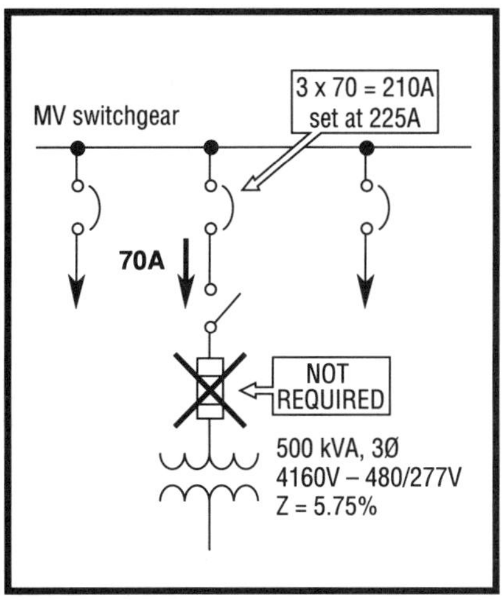

Fig. 5.4. Primary feeder O.C. device can serve as protection for primary of transformer operating at over 600V.

If, on the other hand, the overcurrent protection for the motor circuit was provided by a set of dual-element time-delay fuses, a different set of calculations would be necessary. The total of the maximum permitted rating or settings of the secondary protective devices remains 750A. The maximum allowable rating of the fuses protecting the motor branch circuit is 175% of the motor flc. The calculated maximum rating is 168A. Interpretation of the NEC rules says that in such a case, the next *lower* standard fuse must be tried, and only if that proves to be insufficient to permit motor starting can the next *higher* fuse rating be used. So, 150A fuses are to be used, or if they do not permit motor starting, then 175A fuses are to be used. In either event, the total of the settings and ratings would not exceed the 750A guideline and this design is permissible.

Supervised installations. More liberal rules for overcurrent protection of power transformers can be applied in a facility where the conditions of maintenance and supervision assure that only qualified persons will monitor and service the installation. When designing a new facility, these more liberal NEC rules should be applied when applicable because they can provide considerable savings in equipment.

An example of this less stringent approach (widely applied in industrial locations) is that no individual overcurrent device is required on the primary side of the transformer if the overcurrent device protecting the primary feeder to the transformer meets a 250% for fuses or 300% for circuit breakers limitation. This is illustrated in **Fig. 5.4**. For a more in-depth discussion of the applicable rules, see *Understanding NE Code Rules on Transformers*.

TRANSFORMERS 600V, NOMINAL, OR LESS

This set of rules applies to transformers connected to a system operating at a voltage of 600V or less. The general rule that applies to the sizing of the overcurrent device on the primary of these transformers is that it must be set or rated at not more than 125% of the rated primary current of the transformer.

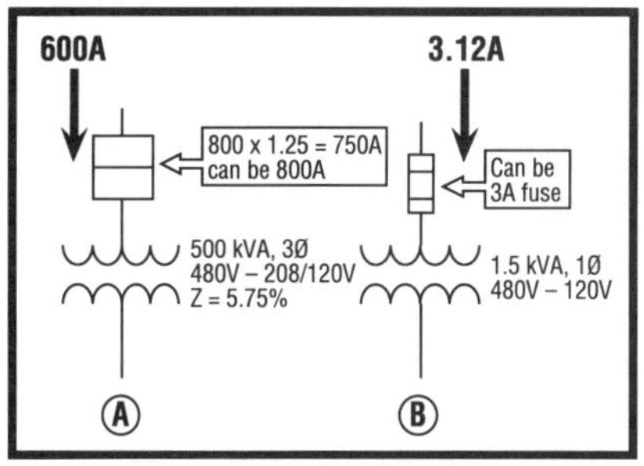

Fig. 5.5. Sizing the primary O.C. protective device of transformer connected to a source operating at 600V or less is shown in examples **(A)** and **(B)**.

There are several exceptions to this rule. The first is dependent upon the rated primary current of the transformer.

- If it is 9A or more, and the 125% does not correspond to a standard rating of a fuse of nonadjustable circuit breaker, the next higher standard rating is permitted.
- If it is less than 9A, the overcurrent device can be rated or set at not more than 167% of the primary current.
- If it is less than 2A, the overcurrent device can be rated or set at not more than 300% of the primary current.

EXAMPLE: The transformer in **Fig. 5.5A** has a rated primary current of 600A, while the one shown in **Fig. 5.5B** has a rated primary current of 3.12A. Size each of the overcurrent protective devices for the transformer primary.

ANSWER: The calculation for the circuit breaker to protect the 500kVA transformer having a 600A primary current is:

$$600 \times 1.25 = 750\text{A}.$$

The exception states that if the current is over 9A, and it is not listed as a standard size, then the next standard size can be used. Thus the CB can be sized at 800A.

The calculation for the fuse to protect the smaller transformer is:

3.12 x 1.67 = 5.2A.

The wording of the exception for a transformer with this level of primary current says that it is to be rated at *not over* 167% of the rated primary current. That means that a 5A fuse is the maximum size allowed. According the NEC, however, *standard* ratings for fuses in that range are 3A and 6A. This would seem to limit the rating of the fuse to 3A. Another code provision, however, states that it is not the intent of the listing to prevent the use of nonstandard CBs or fuses. Thus, a 5A fuse could be used if available.

A transformer whose primary is connected to a system operating at 600V, nominal, or less, is not required to have an overcurrent protective device on its primary if:

- it has an overcurrent device on its secondary side rated or set at not more than 125% of the rated secondary current of the transformer; and
- the overcurrent device protecting the feeder to the transformer primary is rated or set at not more than 250% of the rated primary current of the transformer.

An exception to this rule says that where the rated secondary current of the transformer is 9A or more, and the maximum size of overcurrent device calculated at 125% does not correspond to a standard size listed, the next higher standard size can be used.

Where the rated secondary current is less than 9A, an overcurrent device rated or set at not more than 167% of the rated secondary current is permitted.

If the transformer is provided by the manufacturer with coordinated thermal overload protection, and it is arranged to interrupt the primary current, the individual protective device on the primary side is not required provided that the overcurrent device protecting the feeder to the transformer primary is set at not more than:

- 6 times the rated primary current

Control circuit conductor size, AWG	O.C. device rating	
	Copper	Alum.
18	7	—
16	10	—
14	15	—
12	20	15
10	30	25

Fig. 5.6. Maximum amp rating of overcurrent protective device.

for a transformer with not more than 6% impedance; or

- 4 times the rated primary current for a transformer with an impedance over 6% but not more than 10%.

CONTROL-CIRCUIT TRANSFORMERS

Transformers used in control circuits must have overcurrent protection similar to that required by the rules applying to power transformers. There are, however, specific rules that apply in addition to those for the larger power transformers.

In motor control circuits, both the control-circuit wiring as well as the transformer itself must be protected. In some cases, the overcurrent protection for the transformer is also permitted to protect the secondary wiring. This permission is limited to transformers with 2-wire secondaries. A 3-wire secondary (say 240/120V) would not qualify for these provisions.

In order for this rule for transformer primary-side protection to apply, the rating of the overcurrent protective device for the transformer must not be greater than the maximum permitted overcurrent device for the size conductor used in the control circuit (see **Fig. 5.6**), times the secondary-to-primary voltage ratio of the transformer.

EXAMPLE: Assume that in **Fig. 5.7**, No. 14, 75°C copper conductors are used on the secondary of a 0.5 kVA control-circuit transformer that reduces the 480V operating voltage of the motor to a control voltage level of 120V. What is the maximum fuse size on the primary side of the transformer that this code section permits to protect the secondary conductors?

ANSWER: For a No. 14 conductor, the size of overcurrent protective device permitted for the conductors is 15A. The calculation is:

15 x (120 ÷ 480) = 3.75A.

This answer covers only the ability of these primary overcurrent devices

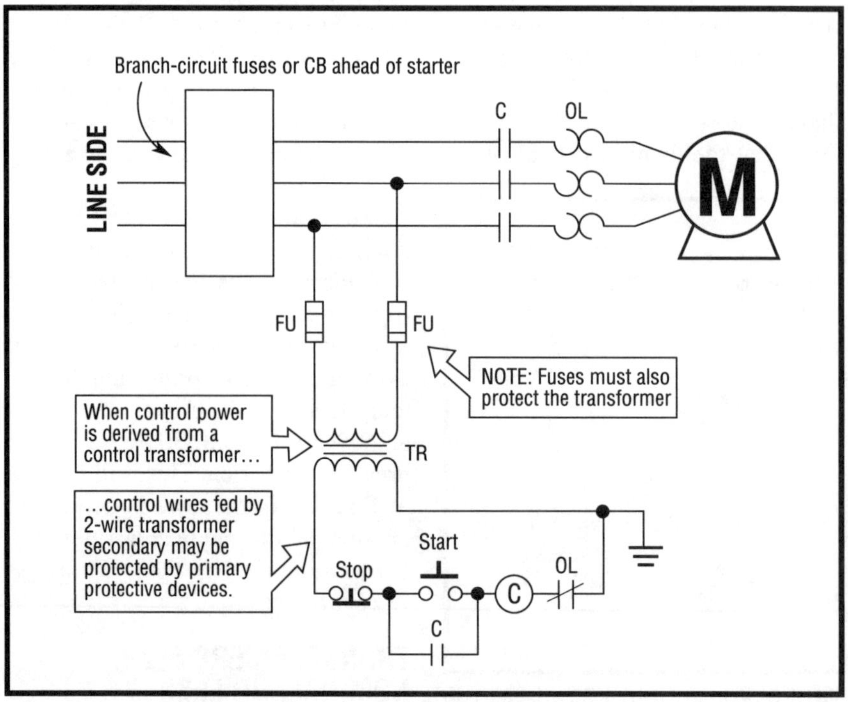

Fig. 5.7. Primary protection may be used for secondary control wires from a control transformer supplying coil current.

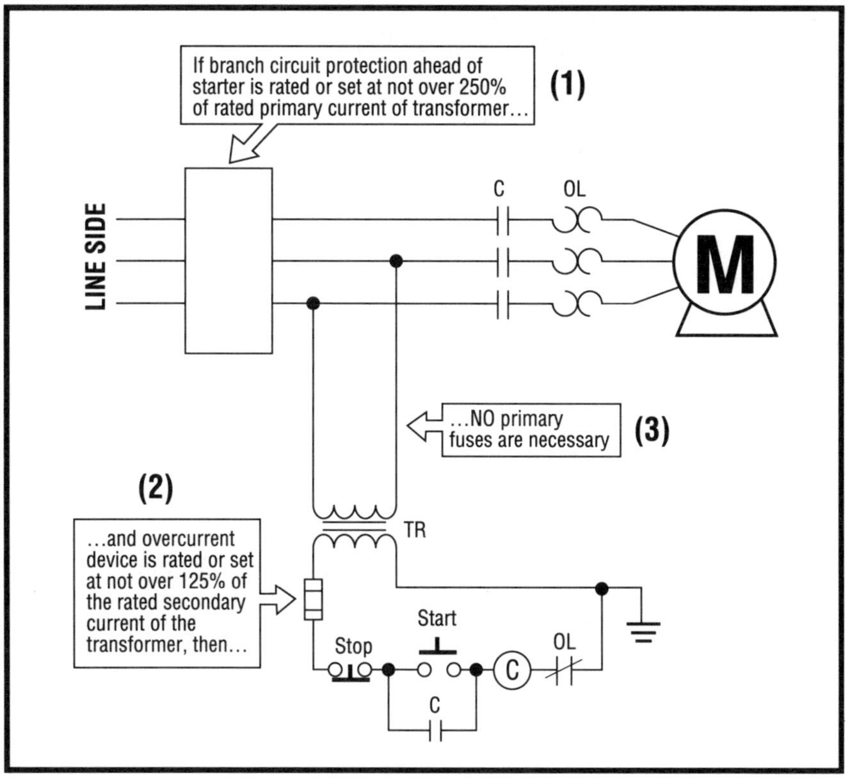

Fig. 5.8. Overcurrent protection on the primary of a control-circuit transformer can also be omitted provided these conditions are met.

to protect the secondary wiring. The general rule is that motor control circuit transformers must be protected in accordance with the previously discussed rules for power transformers. So it is necessary to see if these fuses will also protect the transformer.

In the example, a 480V primary voltage was indicated. The general rule that applies to the sizing of the overcurrent device on the primary of this transformer is that it must be set or rated at not more than 125% of the rated primary current of the transformer, which for a 0.5kVA transformer at 480V is 1.04A. Thus: 1.04 x 1.25 = 1.30A. Fuses sized by this rule to protect the secondary wiring, in this case, would be too large to protect the control transformer. But recall that if the primary current of the transformer is less than 2A, the overcurrent device can be rated or set at not more than 300% of the primary current. The calculation will be:

$$1.04 \times 3 = 3.12\text{A}.$$

A set of 3A fuses would therefore adequately protect both the transformer and the secondary conductors.

The example only covers situations where the transformer has a two-wire secondary. Where the control transformer has a 120/240V secondary, the rules for protecting power transformers apply; the control circuit wiring cannot be protected by the primary overcurrent device.

Overcurrent devices protecting the control circuit conductors can also serve to protect a single-phase, 3-wire control transformer, thus doing away with protective devices on the primary of the transformer. As shown in **Fig. 5.8**), no overcurrent protective device is required on its primary if:

- the overcurrent device on its secondary side is rated or set at not more than 125% of the rated secondary current of the transformer; and
- the overcurrent device protecting the feeder to the transformer primary is rated or set at not more than 250% of the rated primary current of the transformer.

Where the rated secondary current of the transformer is 9A or more, and the maximum size of overcurrent device calculated at 125% does not correspond to a standard size, the next highest standard size can be used. Where the rated secondary current is less than 9A, an overcurrent device rated or set at not more than 167% of the rated secondary current is permitted.

There are other rules that apply to overcurrent protection of control-circuit transformers.

- A control transformer rated less than 50VA that is an integral part of a starter and is mounted within the starter enclosure, can be protected by primary overcurrent devices, impedance limiting means, or other inherent protective means.
- Where the control circuit transformer rated primary current is less than 2A, an overcurrent device rated or set at not less than 500% of the rated primary current is permitted in the primary circuit.
- Other forms of overcurrent protection provided by approved means are allowed.
- Overcurrent protection is not required for a control transformer per the general rule where it supplies a Class 1 power-limited circuit, or Class 2 or Class 3 remote-control circuit.

FIRE-PUMP INSTALLATIONS

When the transformer is dedicated to supplying power to a fire pump installation, secondary overcurrent protection is not required. Failure of the fire pump to start and run when required is a greater potential hazard than is the eventual burning out of the transformer.

Primary protection, however, is required. The overcurrent protective device setting or rating must be sufficient to indefinitely provide the transformer secondary current required to carry the sum of the locked-rotor currents of the fire pump motor(s) and the associated fire pump accessory equipment (such as jockey pumps, controls, alarms, etc.).

OTHER CONSIDERATIONS

Control circuits associated with motor controls can be extremely complex and vary greatly with the application. Besides the items covered previously, there are other code requirements for control-circuit transformers that must be adhered to.

Where a transformer is used to obtain a reduced voltage for the motor control circuit and is located in the controller enclosure, the transformer must be connected to the load side of the disconnecting means for the motor control circuit.

AC systems operating at 50 to 1000V are required to be grounded where the system can be grounded so that the maximum voltage to ground on the ungrounded conductors does not exceed 150V. To meet this requirement, any 120V or less, 2-wire circuit derived from a control transformer must normally have one of its secondary-side conductors grounded.

An exception permits ungrounded control circuits under certain specified conditions. A control circuit may be operated ungrounded when all of the following conditions exist.

- The circuit is derived from a transformer that has a primary rating of less than 1000V.
- Supervision will assure that only persons qualified in electrical work will maintain and service the control circuits.
- There is a need for preventing circuit opening on ground fault, such as for safety or for operating reliability.
- Some type of ground detector is used on the ungrounded system to alert personnel to the presence of a ground fault, enabling them to clear the ground fault during normal system maintenance.

According to the basic rule, when a separately-derived AC system has one of its secondary conductors grounded, it is necessary to bond the grounded conductor to the metal case of the transformer and run a grounding electrode conductor from the grounded transformer secondary terminal to nearby grounded building steel or a grounded metal water pipe. But an exception exempts small control transformers from the basic requirement for a grounding electrode conductor run.

A Class 1, Class 2, or Class 3 circuit derived from a transformer rated at not over 1kVA, however, has to have a grounded secondary conductor bonded to the metal case of the transformer using a bonding conductor not smaller than No. 14 copper or No. 12 aluminum. No grounding electrode conductor is needed provided that the metal transformer case itself is properly grounded.

6

TYPICAL APPLICATION CALCULATIONS

There are many considerations that go into fitting a transformer into the overall design of a power distribution system. The actual design of the transformer itself is very specialized and is beyond the scope of this book. This task is performed by highly skilled professionals, increasingly with the aid of computers.

Most of the application-type calculations are relatively simple and are not beyond the abilities of the average person in the electrical field. This chapter will concentrate on reviewing the basic steps involved.

The easiest way to approach the subject is by first going through a typical example of a transformer application and the steps involved.

APPLYING A POWER TRANSFORMER

A new office area will contain desktop computers, task lights, copy machines, and other small appliances, all operating at 120V. The office HVAC is to be supplied at 480V, and general overhead lighting is to be provided by fluorescent lighting at 277V. A dry-type distribution transformer is to installed to provide power for the 120V loads (**Fig. 6.1**).

Sizing the transformer. Load requirements for the proposed 120V equipment is estimated to be l8kVA. Allowing 20% for future growth brings this value up to 21.6kVA, too high a load for a standard size 15kVA transformer. As a result, the next larger size, 30kVA, is selected.

The unit selected is a conventional 480 to 120/208V, 3-phase dry-type transformer without factory-equipped thermal overload protection to interrupt the primary current. This is an important condition that must be known when selecting transformer protection. The NEC allows the overcurrent protective device on the primary of the transformer to be eliminated where this type of protection is provided by the manufacturer if the primary feeder overcurrent device is rated or set at a value of not more than:

- six times the rated current of a transformer having not more than 6% impedance; or
- four times the rated current of a transformer having more than 6% but not more than 10% impedance.

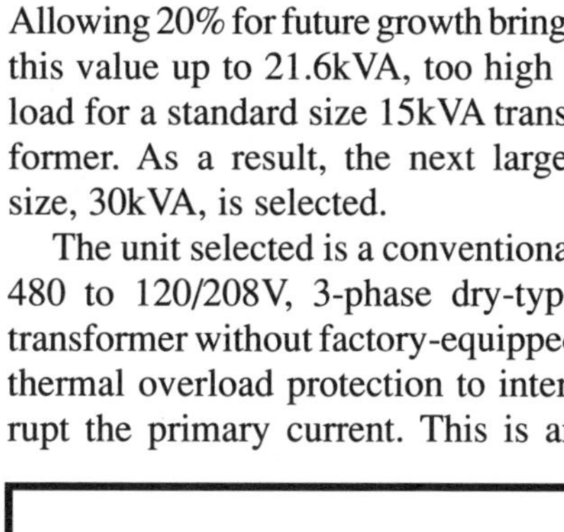

Fig. 6.1. Sizing overcurrent protective devices for a dry-type transformer.

Primary overcurrent protection. Since the thermal overload protection option has not been selected, a primary overcurrent protective device seems to be required. The applicable NEC rule [Sec. 450-3(b)(2)], however, allows the primary overcurrent protective device to eliminated if:

- the overcurrent device on the secondary is rated or set at not more than 125% of the rated secondary current of the transformer; and
- the primary feeder overcurrent device is rated or set at not more than 250% of the rated primary current of the transformer.

The primary protection for this transformer is proposed to be a 60A, 3-pole breaker in an existing 480Y/277V panelboard. The rated primary current of the transformer is 36A. The calculation is:

$$36 \times 2.5 = 90A.$$

The 60A breaker, therefore, is permitted to serve as the overcurrent protection for the transformer primary.

A check should be made at this time to assure against possible tripping of this breaker. The full-load primary current value should be increased by at least 25% to allow the transformer magnetizing inrush current spike to pass without tripping the transformer primary circuit breaker. Thus:

$$36 \times 1.25 = 45A$$

This is less than the existing 60A circuit breaker that is to be used; the 60A device is suitable.

Secondary overcurrent protection. Since the 120V receptacles are to be fed from a new 100A, 208Y/120V panelboard, it is efficient as well as convenient to have a main circuit breaker installed that will serve as the panelboard main disconnect and the transformer secondary overcurrent protective device. According to NEC Article 384, this panelboard is classified as a lighting and appliance branch-circuit panelboard (having more than 10% of its overcurrent devices rated at 30A or less). These types of panelboards must have individual overcurrent protection on their supply side.

As was noted previously, the secondary overcurrent device can serve as the transformer protection provided it is rated or set at not more than 125% of the rated secondary transformer current. Generally the rated full-load secondary current of the transformer is based on information from a catalog. If not available, however, it can be obtained from the equation:

$$I = (kVA \times 1000) \div (V \times 1.73)$$

where:

I = secondary full-load current
kVA = kVA rating of the transformer
V = line-to-line voltage.

In this case, therefore:

$$I = (30 \times 1000) \div (208 \times 1.73) = 83.4A.$$

The secondary overcurrent protective device may be set at not more than:

$$83.4A \times 1.25 = 104A.$$

This rating is not available, so a rule in Article 450 allows the use of the next higher standard rating. This is given in Article 240 as 110A.

Protecting the panelboard. It may seem that this part of the design is finished; but it is not. The rating of the overcurrent device protecting the panelboard is not allowed to exceed the rating of the panelboard.

If the 110A circuit breaker is to be used, it will not properly protect a 100A panelboard. There are two choices; the first being to increase the panel rating to equal or exceed 110A. Depending on the brand of equipment chosen, this would require 125A, 150A, or 200A busing. This is a needless expense, because the additional 10A is not really usable. Remember, it is not permissible to intentionally overload a transformer, and the rated secondary current is only 83A!

The second approach is to design the job around a 100A rated panel with a 100A main breaker. Such a breaker will easily protect both its panel and the transformer. Always remember, however, that the connected load must never exceed 83A that represent the transformer's full-load current rating.

Although not required by the NEC, a further degree of conservatism could be added to the design by sizing the panelboard's main breaker at the next lower size from the transformer rating, or 80A. This value will also allow for the full design loading, even if it is continuous. For a 21.6kVA design load, the current draw is:

$$I = (21.6 \times 1000) \div (208 \times 1.73) = 60A.$$

Overcurrent protective devices normally are sized at 125% of the continuous load, plus any noncontinuous load, if any. Although general purpose office type loads are usually classified as noncontinuous, in this case even if all the load were continuous, the conservative selection is adequate:

$$60 \times 1.25 = 75A;$$

an 80A circuit breaker is adequate.

Sizing conductors. The conductors connecting the transformer secondary to the panelboard main breaker must be selected per Table 310-16, using the 60°C rating because circuit breaker terminals in this ampere range are rated for 60°C. If, however, both the transformer terminals and circuit breaker terminals are rated higher (75° or 90° C), then the lowest of these marked temperature current ratings may be used. To use these ratings, the panelboard as well as the circuit breaker must be marked for the higher-rated terminations.

Because there are four wires in the transformer secondary raceway, the possibility of derating per the Notes to Ampacity Tables must be considered. General office type loads, especially in the past, have not been derated in this case because, under Note 10(a), the neutral would not be included as a current carrying conductor. In this example, a worst case assumption that a majority of the load is harmonic in character will be made. (The effect of harmonics is discussed in Chapter 7 of this book.) The neutral, therefore, will be classified as current carrying, in accordance with Note 10(c).

This will mean that the ampacity of the wires must be adjusted to 80% of their Table 310-16 ampacity as per Note 8(a). Thus, for an 80A circuit breaker, the 60°C wires must be sized at 80 ÷ 0.8 = 100A. This is not listed size, so the next larger size is No. 1 copper wire (rated at 110A).

If the terminations were all rated for 75°C, then No. 3 copper could have been used.

Other details. The minimum size raceway is $1^1/_2$ inches (from Table 3 of

NEC Chapter 9) for four No. 1 conductors in a raceway.

The transformer neutral wire must be connected to ground by a grounding electrode conductor per Sec 250-26(b), as this arrangement is now a "separately derived alternating-current system," as described in Sec. 250-5(b)(2). A No. 6 AWG (minimum) copper wire should be used as per Table 250-94.

There will be only three wires in the primary conduit. These wires (per Table 310-16, rated for 60A protection and considered 55A, 60ºC conductors for termination purposes), could be No. 6 copper (THHN) in a ¾ in. conduit.

APPLYING AN AUTOTRANSFORMER

Use of autotransformers has greatly increased recently for applications in the 600V or less class. Air-conditioning, motors, and electric-heating loads often require 240V for operation. Facilities having 208Y/120V electrical systems often provide an autotransformer that bucks (decreases) or boosts (increases) the available voltage to satisfy operational requirements. The transformer is usually delivered as two 2-winding (insulating) transformers that are field connected as an autotransformer. These connections must be made so as to satisfy the NEC rules for overcurrent protection of autotransformers.

Note that the NEC requires that an insulated transformer can be reconnected for use as an autotransformer to produce a higher-than-rated secondary voltage only if it is so labeled by the manufacturer. While they may look like ordinary 2-winding transformers, the secondary windings have sufficient electrical insulation to allow them to be connected to the same supply lines as the primary windings.

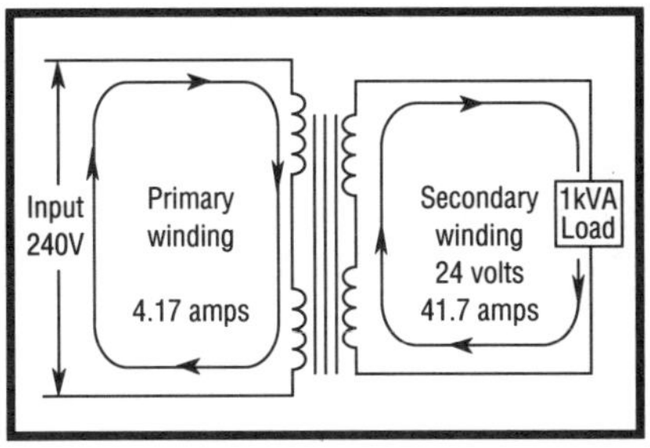

Fig. 6.2. Typical 1kVA boost-buck transformer connected as a standard insulating-type transformer.

An understanding of the operational differences between the "normal" and the autotransformer configuration will ensure the proper application of the required overcurrent protective devices.

Autotransformer rating. For this discussion, a 1kVA, 240/120V to 24/12V transformer will be used as an example. First consider it connected as a standard transformer (**Fig. 6.2**). All the power delivered to the load is transformed by magnetic coupling from the primary to the secondary windings.

The current flowing in the primary and secondary windings equals the VA rating divided by the rated voltage:

$$1000VA \div 240V = 4.17A \text{ (primary)}$$
$$1000VA \div 24V = 41.7A \text{ (secondary).}$$

The identical transformer, shown in **Fig. 6.3** connected as an boost auto transformer (additive polarities), has a voltage available to the load of:

$$240V + 24V = 264V$$

but the load current remains the same at 41.7A, and now the rating of the transformer is:

$$264V \times 41.7A = 11{,}000VA = 11kVA!$$

Supply line current is:

$$11{,}000VA \div 240V = 45.87A.$$

The current in the primary, however, does not exceed the rating of the winding because the supply line and load currents flow in opposite directions. Current in the primary winding is:

$$I_p = 45.87A - 41.7A = 4.17A.$$

A 1kVA transformer in boost-buck service can, thus, be applied to an 11kVA load. By connecting the two secondary windings in parallel rather than in series, a rating of 21kVA can be achieved with the same transformer!

Overcurrent protection. Requirements for the single-phase, 120/240V to 12/24V, insulated transformer rated 1kVA will be analyzed with the unit connected as an autotransformer.

The insulated transformer has been field-connected as a boost autotransformer. Even though the currents through the primary and secondary of the transformer remain the same, the secondary current (41.7A) is being drawn directly from the line, as is the 4.17A primary current. This autotransformer current is known as the *input full-load current*. It is the determining factor in the selection of overcurrent protection for field-connected autotransformers.

The input full-load current in this example has been calculated at 45.87A (say 46A). Now the rules of NEC Article 450 for autotransformers operating at 600V, nominal, or less apply. As before, an overcurrent device of not more than 125% of the input full-load current may be used. Therefore, multiply the input full-load current by 125%:

$$46A \times 1.25 = 57.5A.$$

An overcurrent device rated not more than 57.5A should be used. This is not a standard size device, therefore the next-higher standard ampere-rated device would be permitted. A 60A rated or set device can be used.

If on the other hand, the autotransformer is to be used to buck the voltage from a higher to a lower voltage, the higher voltage would be the input that must contain overcurrent protec-

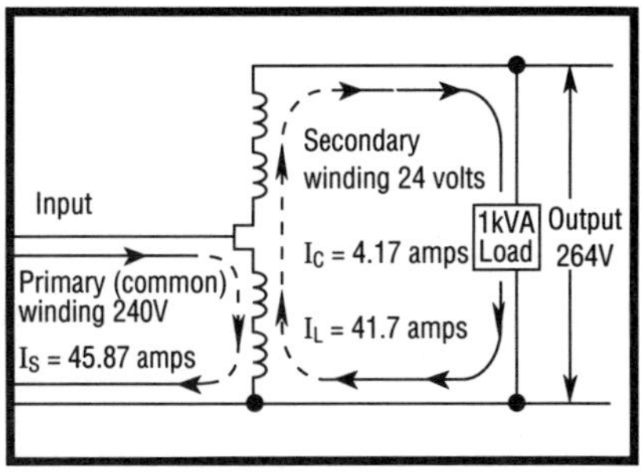

Fig. 6.3. The same 1kVA transformer reconnected as an autotransformer in a voltage boost application. I_s = supply line current; I_L = load current; I_c = common (primary) winding current.

tion. The calculation could result in a different size device.

APPLYING GROUNDING AUTOTRANSFORMERS

A grounding autotransformer is a specialized form of autotransformer. Its function is to derive a neutral from an ungrounded power source. The neutral is then used in several different ways. The transformers covered by this example are zig-zag types installed in 3-phase, 3-wire ungrounded systems to create a 3-phase, 4-wire distribution system, or to provide a neutral reference for grounding purposes.

Creating a 3-phase, 4-wire system. A grounding autotransformer can be connected to a 3-phase, 3-wire system to create a system from which phase-to-neutral loads can be supplied, as shown in **Fig. 6.4**. Overcurrent protection provided for the autotransformer must be of the type that will open the main switch or overcurrent device protecting the 3-phase, 4-wire system if the load on the autotransformer exceeds 125% of its continuous current per phase or neutral rating. Delayed tripping by this device is permitted to allow the branch circuit or main overcurrent protective devices to operate first if the problem is a fault further in the distribution system.

Further relaying must be provided for the autotransformer to cause the opening of the main switch or overcurrent device for the 3-phase, 4-wire system if single-phasing or an internal fault occurs. This can be accomplished by the use of two subtractively-connected donut-type (window-type) CTs installed to sense and signal when an unbalance occurs in the line current to the autotransformer of 50% or more of rated current.

The autotransformer must have a continuous neutral current rating sufficient to handle the maximum possible neutral unbalanced load current of the 4-wire system.

EXAMPLE: An existing 1500kVA, 3-phase transformer with a 480V delta-connected secondary is to carry 600kVA of newly-installed 277V fluorescent lighting, which is evenly balanced between the phases. What size zig-zag autotransformer is required to generate the neutral and at what value must the overcurrent and fault relaying trip the main overcurrent device?

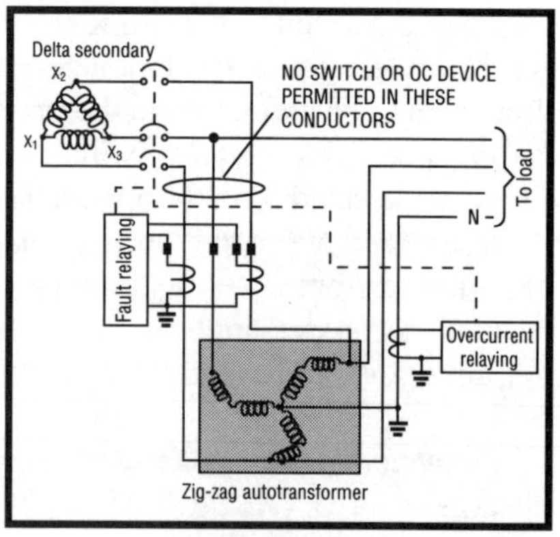

Fig. 6.4. A zig-zag grounding autotransformer used to create a fourth grounded conductor on a 3-phase, 3-wire system.

ANSWER: The maximum unbalanced current that could flow is

200kVA ÷ 277V = 722A.

Because of the nature of the zig-zag connection (see Chapter 3), 1/3 of the neutral current will flow through each leg of the autotransformer. For maximum unbalance, the current through each will be:

722 ÷ 3 = 240.7A.

Fig. 6.5 shows the maximum continuous kVA or ampere load per phase that can be carried by zig-zag auto transformers. The nearest to meeting the specification are three 25kVA units that can sustain a maximum continuous amp load per phase of 312A.

Overcurrent protection is sized at a maximum current in the neutral:

3 x 312 x 1.25 = 1179A.

Transformer fault protection is also required. The amount of unbalance in the line current beyond which tripping of the main protective device would be required is:

312 x 0.5 = 156A.

USING TIME-CURRENT CHARACTERISTIC CURVES

There are many situations where a machine tool transformer (control transformer) is used to reduce the line voltage to the operating voltage of a motor's control circuit. Where a motor control-circuit transformer is used, the NEC requires overcurrent protection. The code allows the rating of the protective device for the primary to be as much as 167% of the primary full-load current for transformers whose primary current is less than 9A; 500% of primary full-load current where primary current is less than 2A; and permits transformers 50VA or less to have no overcurrent protection.

It might seem from these rules that smaller transformers are less likely to short-circuit or be overloaded. This is not the case. The reason for the flexibility is to prevent nuisance interruptions caused by the inrush current associated with control-circuit transformers, which can have peak values from 4 to 26 times the operat-

Nameplate KVA Each	No. Required	Three Phase KVA	Max. Continuous Amp. Load Per Phase (277 Volts)
1.0	3	10.8	12.50
1.5	3	15.6	18.75
2.0	3	20.7	25.00
3.0	3	31.2	37.50
5.0	3	51.9	62.50
7.5	3	78.0	93.50
10.0	3	103.8	125.00
15.0	3	156.0	187.50
25.0	3	259.5	312.00
37.5	3	390.0	468.00
50.0	3	519.0	625.00
75.0	3	780.0	935.00
100.0	3	1038.0	1250.00
167.0	3	1734.0	2085.00

Fig. 6.5. Typical zig-zag autotransformer data table.

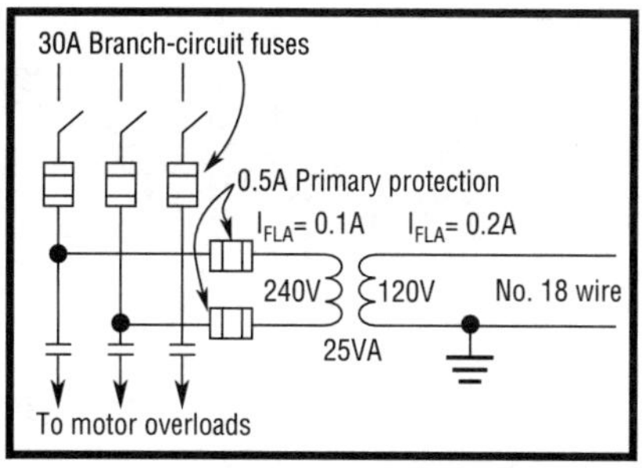

Fig. 6.6. Even though 30A branch circuit fuses are permitted by the NEC to protect the control transformer, the excessive heat that will be generated before these fuses part could cause a transformer fire and damage to secondary wiring. A better idea is to supply the control transformer with its own protection, here shown as 0.5A fuses.

ing current of the transformers. It is often advisable to provide protection even though not required by the NEC (which is a *minimum* requirement.

Consider the use of low-amperage fuses as an overcurrent device to protect the 25VA control transformer (**Fig. 6.6**). The primary full-load current is 0.1A. It is permissible in accordance with Exception No.2 to Sec. 430-72(c) to use overcurrent devices rated 500% of the primary full-load current. This results in the selection of $^1/_2$A standard fuses. Assume for this example that the inrush equals 26 times the primary full-load current. Upon application of power to the branch circuit, the control transformer overcurrent device would be subjected to 2.6A peak current for the first half-cycle (.0083 sec).

Time-current characteristic curves, such as the one shown in **Fig. 6.7**, are used to find out if the selected overcurrent devices can withstand the inrush for $^1/_2$ cycle. Because the ampere ratings on these curves are rms values, the 2.6A peak current must be converted to an rms value by dividing by 1.414. The calculated rms value in this example is about 1.83A. From the time-current characteristics, these fuses can withstand about 3.5A rms for 0.01seconds, which should not result in nuisance interruptions on inrush. These devices will provide short-circuit protection for the control transformer, but an overload condition where the primary current is five times the normal current (0.5A) would not interrupt the fuse. One of 0.8A would take 0.5 seconds to blow the fuse and could still cause damage before the overcurrent protection opens!

Potential for damage due to overload can be minimized by reducing the rating of these fuses to some value less than 500% of primary full-load current. The general rule is that primary overcurrent protection must not be rated more than 125% of primary full-load current. The primary full-load current is 0.1A x 1.25 = 0.125A ($^1/_8$ A). The characteristics curve shows the fuse provides better overload protection. For example, if the primary current reached 0.3A, the fuse would blow in 0.05 seconds.

But the characteristic curve shows that the $^1/_8$ amp fuse would not be able

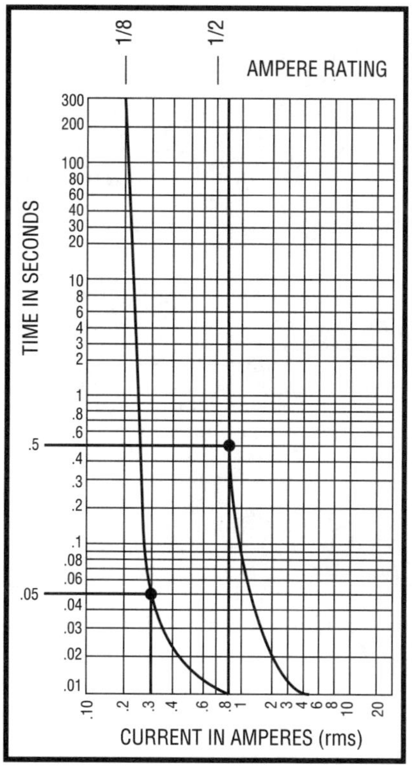

Fig. 6.7. Time-current characteristic curves for 0.50A and 0.125A fuses.

to withstand the 1.83A inrush current. Thus, because of the inrush, the overcurrent devices must be sized at 500% of primary full-load current at the sacrifice of better overload protection.

This does not have to be the case. Although the inrush is greater than the operating current, it only lasts for $^1/_2$ cycle, and therein lies the solution to providing functional short-circuit pro-

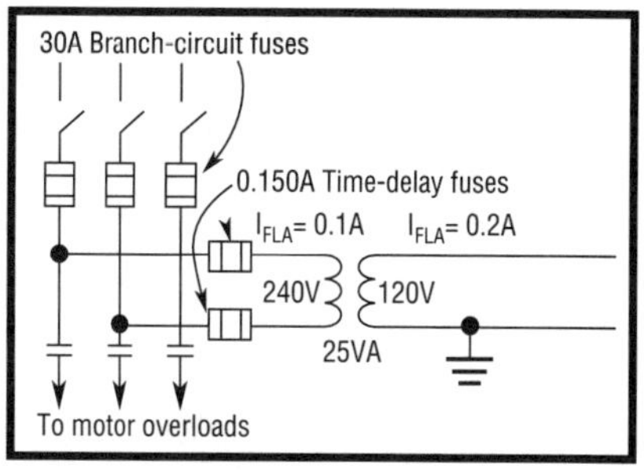

Fig. 6.8. Control-circuit transformer protected with time-delay fuses.

tection and a higher level of overload protection for the control transformer.

The solution is time-delay fuses. In **Fig. 6.8**, 0.15A ($^{15}/_{100}$A) time-delay fuses have been used because fuses of these types do not come in $^1/_8$A sizes. The time-current characteristics curve for this fuse (**Fig. 6.9**) shows that the fuse can withstand about 12A rms for 0.01 seconds, which is greater than the 1.83A rms and would not cause nuisance interruptions. Further, if during an overload the primary current reached 0.3A rms, this fuse would clear the overload in 30 seconds, thereby reducing the risk of heat damage to the control-circuit conductors and control transformer.

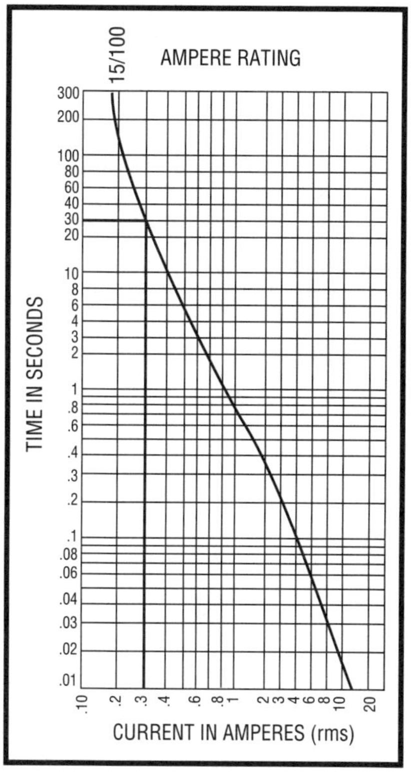

Fig. 6.9. Time-current characteristic curve for 15/100A time-delay fuses.

Computerization of the workplace has drastically affected power distribution systems by greatly increasing the percentage of nonlinear load carried by transformers and other electrical equipment.

NONLINEAR LOADS AND HARMONICS

7

Nonlinear electrical loads such as computers, solid-state power electronics, variable-frequency drives, and others, have expanded rapidly in recent years. This has led to serious and unexpected problems, such as failure of transformers loaded well below their maximum ratings, and severe overheating of full-size neutral conductors. These new loads behave differently from normal linear loads and require a revision of thinking in applying this equipment to avoid failures.

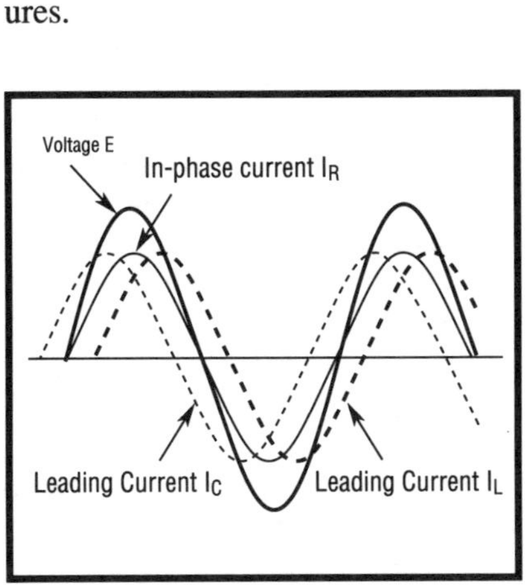

Fig. 7.1. Linear currents: I_R is a pure resistive circuit current; I_L is a partially inductive (lagging) circuit current; and I_C is a partially capacitive (leading) circuit current, which is uncommon.

THE EFFECT OF LOADS

Motors, incandescent lighting, and heating loads are linear in nature. That is, the load impedance is essentially constant regardless of the applied voltage. As seen in **Fig. 7.1**, in AC circuits the current waveform looks the same as that of the voltage, following the way the voltage increases and decreases. The current in these circuits is in phase with the voltage for a resistive circuit, with a power factor (PF) of unity. It lags the voltage by some phase angle for the more typical partially-inductive circuit containing a transformer (with a PF commonly between 0.80 and 0.95), and leads the voltage by some phase angle in a capacitive circuit. For a sinusoidal voltage, the current is also sinusoidal.

Until recently, almost all loads were linear, and those that were not were such a small portion of the total that they had little effect on system. Now electronic loads, such as computers, UPS equipment, and variable-speed motor drives have become a large enough factor and have serious consequences on power distribution systems, including equipment such as transformers.

A nonlinear load is one in which the load current is not proportional to the instantaneous voltage (**Fig. 7.2**). Often, the load current is not continuous. This type of load can be switched on for only part of the cycle, or pulsed. Nonlinear load currents are nonsinusoidal, and even when the source voltage is a clean sine wave, the nonlinear loads will distort that voltage wave, making it nonsinusoidal.

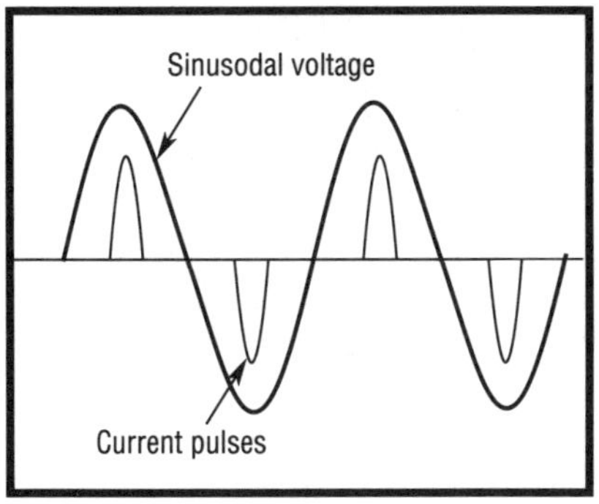

Fig. 7.2. A typical nonlinear load current.

HARMONICS

Any waveshape can be reproduced exactly by adding together a series of sinewaves of particular frequency, amplitude, and timing, although it may require an infinite number of them. Non-sinusoidal waveforms, therefore, consist of (and can be broken down into) some finite number of pure

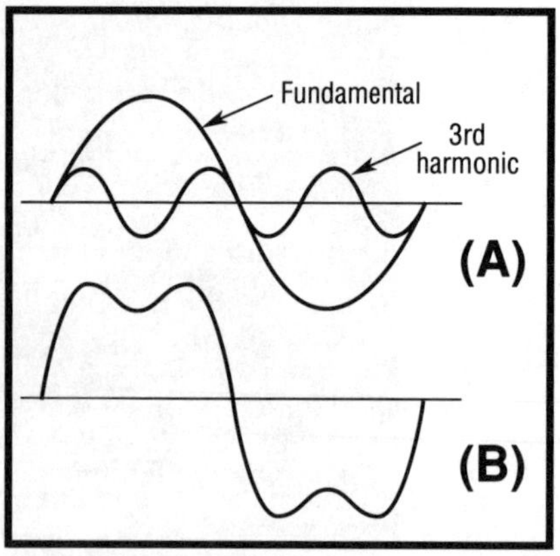

Fig. 7.3. The addition of a 3rd harmonic wave shape to a 60Hz sine wave **(A)**, results in the distorted wave shown in **(B)**.

sinewaves. The most common sinewaves that distort a power system are whole number multiples of the fundamental 60Hz power frequency. These multiples are called harmonics. **Fig. 7.3** shows how the addition of a 3rd harmonic to a sine wave creates a distorted waveshape.

In a 3-phase, 4-wire system, single-phase line-to-neutral load currents flow in each phase conductor and return in the common neutral conductor. The three 60-Hz phase currents are separated by 120° and for balanced 3-phase "linear" loads, they are equal. When they return in the neutral, they cancel each other out, adding up to zero at all points. Therefore, for balanced 3-phase, 60-Hz loads, neutral current is zero.

For 2nd harmonic currents separated by 120°, cancellation in the neutral is also be complete, resulting in zero neutral current. This is true for all *even* harmonics.

For 3rd harmonic currents, the return currents from each of the three phases are in phase in the neutral, so the total 3rd harmonic neutral current is the *arithmetic sum* of the three individual 3rd harmonic phase currents. This is also true for all *odd* multiples of the 3rd harmonic (9th, 15th, 21st, and so on).

Other odd harmonics (5th, 7th, 11th, 13th, and so on) add in the neutral, but the total neutral-harmonic current is somewhat less than the arithmetic sum of the three harmonic phase currents. Mathematically, the total is the *vector sum* of the three currents. The phase angles between the three phase currents result in partial addition and partial cancellation.

The theoretical maximum neutral current with harmonics is at least 1.73 and perhaps as much as 3.0 times the phase current.

For pulsed loads, the pulses can occur in each phase at a different time. They will return in the common neutral, but they will be separated by time; therefore, there will be no cancellation. If none of the pulses overlap, the neutral current can be three times the phase current.

The lack of neutral current cancellation when nonlinear loads are carried by a transformer is the reason for overheated neutrals.

THE PROBLEM

Ratings of transformers are based on the heating created by load currents of an undistorted 60-Hz sine wave. When the load currents are nonlinear and have a substantial harmonic content, they cause considerably more heating than the same number of amperes of pure sine wave. There are several major reasons for this.

Hysteresis. When steel is magnetized, the minute particles known as magnetic domains (see Chapter 1) reverse direction as the current alternates, and the magnetic polarity also reverses. The magnetizing of the steel is not 100% efficient, since energy is required to overcome the friction of the magnetic domains. This creates hysteresis losses, resulting in heat in the steel laminations of the core. The hysteresis losses are greater for a given rms current at the higher-frequency harmonics, where the magnetic reversals are more rapid than at the fundamental 60 Hz.

Eddy currents. Alternating magnetic fields also induce currents into the steel laminations when the changing magnetic flux cuts through a conductor. These "eddy currents" flow through the resistance of the steel, generating I^2R eddy-current heating losses. Because of the higher frequencies, eddy-current losses are considerably greater for harmonic currents than they are for the same rms value of 60-Hz current.

Skin effect. A lesser, but still considerable, heating effect at higher frequencies is caused by the "skin effect" in the conductors. Currents at higher frequencies are not distributed evenly through the cross-section of the conductor. The magnetic fields tend to force the current flow toward the outside or skin of the conductor. This effect increases as the frequency increases, and also as the magnitude of current increases. At higher frequencies, the center of the conductor carries little or no current. Therefore, the effective cross-section of the conductor is decreased, and its resistance is increased. It behaves as a smaller conductor of lower ampacity. As a result, a given current at harmonic frequencies causes more conductor heating than the same current at 60 Hz.

The result of hysteresis, eddy current, and skin effect is that the transformer carrying no more than its full-rated rms current, but supplying nonlinear loads with a high harmonic content, will overheat, sometimes to the point of failure. As the harmonic currents are drawn by the loads, they act on the impedance of the source, causing harmonic distortion of the source voltage.

EFFECT ON TRANSFORMER OVERCURRENT PROTECTION

When the load on a delta-wye connected transformer is a source of harmonic currents, the third and multiples of the third harmonic will circulate in the delta-connected windings and not appear as input line current to the transformer. For this reason the input line current does not reflect the transformer's total load current, and the kVA input to these transformers seems to be less than the kVA output.

The input circuit breakers should be sized to protect against faults. Since input current is not a measure of the total kVA load in the transformer, the input circuit breakers cannot protect against overload. For this reason, protection must be provided by circuit breakers in the wye-connected secondary output circuits.

DEFEATING HARMONICS

There are several ways in which transformers supplying nonlinear loads can compensate for harmonics. Here are some of the recommendations.

- Derate the transformer.
- Select three-phase transformers with low internal impedance, preferably in the 3% to 5% range, and always of the delta primary, wye secondary configuration. The transformer should be of three-legged core construction rather than three single-phase transformers or an open-delta arrangement. An open delta transformer is not recommended because it fails to provide a low-impedance path for the third harmonic current, and does not provide the power factor correction that a delta-connected winding can supply.
- As an alternative to changing a building's transformer and its wiring, install one or more delta/wye transformers designed for rectifier loads.
- Although properly selected power factor correcting capacitors connected on the load side of the transformer may reduce the harmonic currents that the transformer may otherwise have to carry, this is not a recommended practice. Power system resonances and unexpected high harmonic current may occur unless the system has been specifically designed to avoid them. Also, each system reconfiguration will require reevaluation of the design.
- A periodic inspection should be made and should include: measurement of phase and neutral currents, and the temperature of transformers and their connections. A portable infrared temperature detector is convenient and effective for this purpose. If high temperatures are detected (above 50°C), determine if excessive currents or loose connections are the cause.

TRANSFORMER DERATING FOR PHASE-TO-NEUTRAL LOADS

A method of calculating the transformer harmonic derating factor (THDF) for phase-to-neutral loads has been provided by CBEMA. This method should *not* be used to derate transformers powering phase-to-phase nonlinear loads because it is only applicable in conditions where there is a high 3rd harmonic content with high crest factor.

Using this derating method, it is necessary to measure the true-rms phase current and the instantaneous peak phase current for each phase of the secondary. While the measurements are simple to make, the test equipment must be capable of taking these true-rms type of readings.

To determine the derating factor for the transformer, the instantaneous peak current (I_{peak}) and true-rms current (I_{rms}) measurements are taken for the three phase conductors. If the phases are not balanced, the numbers used should be an average of the three phases. These values can then be used in the following equation:

$$THDF= (1.414 \times I_{rms}) \div I_{peak}.$$

This equation generates a value between 0 and 1.0, although typically between 0.5 and 0.9. (Note that if the phase currents are purely sinusoidal, the instantaneous peaks would be 1.414 times the true-rms value, and the derating factor would be 1.0 (in other words, no derating.)

The transformer rating with harmonics present would then be:

$$kVA_{derated}= kVA_{nameplate} \times THDF.$$

Once the transformer has been derated, the next step is to determine the amount of kVA the transformer is actually delivering and compare that amount to its derated capacity. The actual kVA for a balanced load can be calculated from the true-rms phase-to-phase voltage (V_{rms}) and the true-rms phase current (I_{rms}):

$$kVA_{actual}= (1.732 \times V_{rms} \times I_{rms}) \div 1000$$

If the actual kVA is close to the derated load capacity of the transformer, consider providing more cooling for the transformer. If additional cooling is not practical, loads may need to be shifted to another transformer.

DERATING TRANSFORMERS SUPPLYING PHASE-TO-PHASE LOADS

When applying a transformer to 3-phase loads, its performance capability when subjected to harmonics can be determined by applying the ANSI/IEEE C57.110-1986 standard, *Recommended Practice for Establishing Transformer Capability When Supplying Nonsinusoidal Load Currents.* This standard can be applied through calculations made on a hand-held calculator. Although not as accurate as computer analysis, the relatively simple calculations involved are, nevertheless, conservative enough to be effective.

Losses. When direct current is passed through transformer winding conductors, a simple I^2R loss results, where R is the DC resistance of the winding. When an alternating current of the same magnitude is passed through these same conductors, an additional loss called *stray loss* is produced. Although all this additional loss is eddy-current loss, only that portion in the transformer windings is actually called "eddy-current loss" (P_{EC}); the portion outside the windings is termed "other stray loss" (P_{OSL}). The total load loss in watts (P_{LL}) of a transformer can then be expressed by:

$$P_{LL} = I^2R + P_{EC} + P_{OSL}$$

Eddy-current loss is proportional to the square of the load current and the square of the frequency, thus nonsinusoidal load currents will cause excessive winding loss resulting in an abnormal winding temperature rise. The effect of these currents on "other stray losses" is much less critical. The I^2R loss will also increase because of these harmonic currents, although not to the extent of eddy-current loss. The basic intent of the IEEE standard is to reduce the losses in winding conductors under harmonic load conditions to that level under rated 60-Hz operating conditions.

Preliminary calculations. The IEEE end-user transformer capability calculation is based on data found in a certified transformer test report, such as listed in the appendixes of ANSI/

IEEE C57.12.90-1987 *Liquid-Immersed Distribution, Power, and Regulating Transformers and Guide for Short-Circuit Testing of Distribution and Power Transformers* or ANSI/IEEE C57.12.91-1979, *Standard Test Code for Dry-Type Distribution and Power Transformers*. Most manufacturer test reports are based on, and utilize, these standards.

EXAMPLE OF CALCULATING TRANSFORMER DERATING

The best way to illustrate the calculations required to determine derating according to the standard is by applying it to a problem.

EXAMPLE: Assume a 2500kVA, 12kV delta-480V delta dry-type distribution transformer powering a significant nonlinear industrial load. Determine the transformer's actual capability as a result of this load.

PROCEDURE: The transformer's certified test report indicates the following:

High-voltage winding resistance:
3.8150 ohms at 170°C
Low-voltage winding resistance:
0.00411 ohms at 170°C
Load losses at 170°C: 31,232W

Some basic information must first be extracted from the test report before the actual derating calculation can be made. Look at winding resistance. Check if a note on the certified test report indicates that the resistance values shown are actually the sum of the three phases in series. If so, then a calculation of the individual winding resistance must be made. If the resistance values noted in the test report are not the sum of the three phases in series, no calculation is required and the noted values can be used as is.

Assume that in this case the resistance values were given in series. For delta windings, the individual winding resistance is equal to $^2/_9$ of the 3-phase series resistance. For wye windings, the individual resistance is $^2/_3$ of the 3-phase series resistance. In the example then, the DC resistance measured between two primary terminals (R_1) is:

$$R_1 = {}^2/_9 \times 23.8150 = 0.84778 \text{ ohms}$$

and the DC resistance measured between two secondary terminals (R_2) is:

$$R_2 = {}^2/_9 \times 20.00411 = 0.00091 \text{ ohms.}$$

Another bit of information needed is the rms primary and secondary sinewave line currents of the transformer. By using a basic electrical equation, this value can be calculated. The primary rms sinewave line current under rated frequency and load conditions ($I_{1\text{-}R}$) is calculated by:

$$\begin{aligned} I_{1\text{-}R} &= (\text{kVA})(1000) \div (1.73)(E) \\ &= (2500)(1000) \div (1.73)(12{,}000) \\ &= 120.28\text{A} \end{aligned}$$

For the rms secondary sinewave line current ($I_{2\text{-}R}$), the calculation is:

$$\begin{aligned} I_{2\text{-}R} &= (\text{kVA})(1000) \div (1.73)(E) \\ &= (2500)(1000) \div (1.73)(480) \\ &= 3007\text{A} \end{aligned}$$

Transformer capability calculation — Step 1. Now that the required basic information is on hand, the first step involves calculating the *winding eddy-current loss under rated conditions* ($P_{EC\text{-}R}$). In other words, the loss in watts due to eddy currents that can be expected with the transformer operating within its rated limits while powering linear loads. The required equation is:

$$P_{EC\text{-}R} = P_{LL} - K[(I_{1\text{-}R})^2 R_1 + (I_{2\text{-}R})^2 R_2]$$

Although the equation may look threatening, it simply requires subtracting the I^2R losses for the primary and secondary from the total transformer load loss in order to obtain eddy-current loss in watts.

All the values required have already been calculated or are from the test report. The value for P_{LL} is found in the test report listed as the "load losses at 170°C." The values for R_1, R_2, $I_{1\text{-}R}$, and $I_{2\text{-}R}$ have already been calculated. The values of K are found in ANSI/IEEE C57.12.91-1979; K equals 1 for single-phase transformers, and 1.5 for 3-phase transformers. Substituting these values:

$$\begin{aligned} P_{EC\text{-}R} &= 31{,}232 - 1.5[(120.28)^2 (0.84778) + (3007)^2 (0.00091) \\ &= 31{,}232 - 30{,}740 \\ &= 492\text{W} \end{aligned}$$

Step 2. A calculation is needed to determine the transformer's maximum secondary winding eddy-current loss (Max $P_{EC\text{-}R}$) in *watts per unit* of that winding's I^2R loss. A per-unit number is a pure number with no unit name, such as watts or amps, attached to it. In this Step 2 calculation, watts divided by watts cancels each other and equals 1. The end result is a number multiplied by 1, which equals the number.

This calculation requires selecting between two equations with the selection dependent on the transformer's turns ratios and current rating.

For transformers with turns ratios not exceeding 4:1, the equation is:

$$\text{Max } P_{EC\text{-}R} = 2.4\, P_{EC\text{-}R} \div K(I_{2\text{-}R})^2 R_2$$

For transformers with turns ratios exceeding 4:1, having one or more windings with maximum self-cooled current rating exceeding 1000A, the equation is:

$$\text{Max } P_{EC\text{-}R} = 2.8\, P_{EC\text{-}R} \div K(I_{1\text{-}R})^2 R_2$$

Since the transformer in the example has a turns ratio exceeding 4:1 and the secondary current ($I_{2\text{-}R}$) exceeds 1000A, the second equation must be used. As in the previous calculation, K= 1.5 since the transformer is 3-phase. Substituting the value of $P_{EC\text{-}R}$ obtained from Step 1 and the information obtained from the preliminary calculations:

$$\begin{aligned} \text{Max } P_{EC\text{-}R} &= 2.8\, P_{EC\text{-}R} \text{ watts} \\ &\div K(I_{2\text{-}R})^2 R_2 \text{ watts} \\ &= (2.8)(492) \\ &\div (1.5)(3007)^2(.000913) \\ &= 0.111 \text{ per unit.} \end{aligned}$$

Step 3. At this point, the IEEE calculation makes the assumption that measurements have been made on the electrical system of which the transformer is part. For our example, assume that in the process of verifying the harmonic problem, a printout of rms currents at each of the harmonics was obtained through the use of a harmonic analyzer. To determine the per-unit values of these printed currents, or I_h, simply divide the recorded rms current at harmonic "h" by the rated rms current. Assume that **Table 7.1** lists the values for the transformer in the example.

h (harmonic order)	I_h (rms current per unit)
1	0.978
5	0.171
7	0.108
11	0.044
13	0.028
17	0.015
19	0.010

Table 7.1. Assumed printout values.

With this information, it is possible to calculate f_h, the *harmonic distribution factor* for each harmonic "h." This is simply the harmonic "h" component of current divided by the fundamental (1st harmonic) 60-Hz component of current. For example, to find the harmonic distribution factor for the 5th harmonic (f_5), the following simple division is required.

$$f_5 = I_5 \div I_1 = 0.171 \div 0.978 = 0.175$$

A tabulation like that shown in **Table 7.2** can them be made of the harmonic distribution factors.

The sums of f_h^2 and $f_h^2h^2$ are used in later steps. The sum noted for the column $I_h^2h^2$ actually represents the "k-factor" rating of the transformer. This factor has been devised to rate the intensity of the nonlinear load and its effects on transformer eddy-current losses. UL, per its standards UL 1561 and 1562, will list transformers rated for certain types of harmonic loads, depending on the harmonic content. UL is standardizing on k-factor ratings of 1, 4, 9, 13, 20, 30, 40, and 50. Nonlinear loads with a k-factor greater than 30 are rarely seen.

Step 4. Now it is necessary to calculate the load loss density under rated conditions per unit of rated I^2R loss density (P_{LL-R}). As the other steps have involved calculations on a per-unit basis, the winding loss density is also considered on this basis. The original equation,

$$P_{LL} = I^2R + P_{EC} + P_{OSL}$$

represents the total load loss of a transformer. If looking at this loss on a per unit of rated I^2R loss basis, it would be necessary to divide each variable by I^2R. Also, for the purpose of this calculation, P_{OSL} is assumed to be zero because, relative to eddy currents, the effect of harmonic currents on "other stray loss" is very minor. The IEEE standard is conservative enough to compensate for this assumption. The equation then becomes:

$$P_{LL-R} = 1 + \text{Max } P_{EC-R}$$

Substituting the value of Max P_{EC-R} obtained from Step 2:

$$P_{LL-R} = 1 + 0.111 = 1.111$$

Step 5. Next, determine the maximum permissible rms nonsinusoidal load current *per unit of rated rms load current* (I_{max}). The following equation is used for this calculation.

$$I_{max} = [P_{LL-R} \div 1 + \{(\text{SUM } f_h^2h^2 \div \text{SUM } f_h^2) \text{ Max } P_{EC-R}\}]^{1/2}$$

Don't feel threatened by the apparent complexity of this equation. The terms that need to be inserted have already been calculated, or come from the sums at the bottom of the columns in the tabulation shown in Step 3. The fact that this equation is raised to the ½ power only means taking the square root of the final number within the brackets.

$$\begin{aligned} I_{max} &= [1.111 \div 1 + \{(2.8339 \div 1.0458)(0.111)\}]^{1/2} \\ &= [1.111 \div 1 + (2.710)(0.111)]^{1/2} \\ &= [1.111 \div 1.301]^{1/2} \\ &= [0.854]^{1/2} = 0.92 \end{aligned}$$

In the example, then, the transformer capability with the given nonsinusoidal load current harmonic composition is 92% of its sinusoidal load current capability.

Step 6. To obtain the maximum permissible secondary rms load current ($I_{2\text{-}Rmax}$), the following equation is used.

$$I_{2\text{-}Rmax} = (I_{2\text{-}R})(I_{max})$$

By substituting the result in the preliminary calculation for $I_{2\text{-}R}$ and the value for I_{max} obtained from Step 5, the following value is obtained.

$$I_{2\text{-}Rmax} = (3007)(0.92) = 2766.4\text{A}$$

Step 7. This final step will determine the derated kVA rating ($kVA_{derated}$) and involves the following equation.

$$\begin{aligned} kVA_{derated} &= (1.732)(I_{2\text{-}Rmax})(E) \div 1000 \\ &= (1.732)(2766.4)(480) \div 1000 \\ &= 2299.9\text{kVA} \end{aligned}$$

ANSWER: Based on the IEEE transformer capability calculation, the rating of the 2500kVA transformer used in the example should be re-

h	I_h	I_h^2	f_h	f_h^2	h^2	$(I_h^2)(h^2)$	$(f_h^2)(h^2)$
1	0.978	0.957	1.000	1.0000	1	0.957	1.000
5	0.171	0.029	0.175	0.0306	25	0.725	0.765
7	0.108	0.012	0.110	0.0122	49	0.588	0.598
11	0.044	0.002	0.045	0.0020	121	0.242	0.242
13	0.028	0.00078	0.029	0.0008	169	0.132	0.135
17	0.015	0.00023	0.015	0.0002	289	0.067	0.0578
19	0.010	0.00010	0.010	0.00001	361	0.036	0.0361
		SUM		**1.04581**		**2.747**	**2.8339**

Table 7.2. Tabulation of harmonic distribution factors.

duced to approximately 2300kVA as a result of the nonlinear load it must power.

Some harmonic analyzers automatically calculate the percentage derating required using the IEEE C57.110-1986 method. The obvious question is how does the machine know the values of the primary and secondary winding resistances and transformer load loss. The answer is that it does not. The software program simply bypasses these values and inserts an arbitrary number for the winding eddy-current loss under rated conditions (P_{EC-R}). Because the eddy-current losses at 60 Hz are very small compared to other losses, accuracy of data is required to keep from grossly over- or under-estimating P_{EC-R}. This point should be kept in mind when arbitrary values of P_{EC-R} are used.

TRANSFORMER K-FACTOR RATINGS

Derating the transformer as a method to reduce temperature rise is not necessarily the best approach. By not recognizing that the transformer is deliberately oversized for harmonic conditions, the end-user could load the unit to its nameplate capacity.

Third-harmonic currents circulating in the primary delta winding also cause an increase in flux. This condition coupled with only a small amount of overvoltage could easily cause the core to saturate. If the core saturates, the magnetizing current could become greater than the full-load current and result in the burnout of the primary winding.

Transformers specifically designed for nonlinear loads have an advantage in that they have lower-than-normal flux densities and, therefore, tolerate a condition of high line voltage coupled with circulating harmonic currents.

Listed dry-type general purpose and power transformers are available that are affected by nonsinusoidal currents in accordance with ANSI/IEEE C57.110-1986 standard. Testing is done to determine factors relating to rms current at certain harmonic orders in a specified way that correlated with heating losses. The factors involved in the tests were collectively called the k-factor. The k-factor is a numerical rating for specifying transformers for nonlinear loads.

For a transformer to handle a large k-factor load, several things are done.

- To attenuate higher harmonics, an electrostatic shield is placed between the primary and secondary windings.
- Because the neutral current is considerable greater than the phase current, the neutral terminal is doubled in size for additional customer neutral cables. The user, in turn, must provide larger capacity neutral conductors to the load.
- Oversized primary conductors are used to compensate for circulating harmonic currents.
- For larger size transformers, the transposition of conductors within the turns of the primary delta windings can also result in reduced losses.
- The secondary is also given special consideration. As the frequency increases to 180 Hz (3rd harmonic) and greater, the skin effect (where current begins to travel more on the circumference of the conductor) becomes more pronounced. This is compensated for by providing greater conductor ampacity in the form of windings composed of several smaller sizes of conductor; the circumference of the total conductors is greater.
- Some transformers utilize bare winding conductors as a cost savings. By using insulated conductors where harmonic currents are present, the secondary wye conductor losses can be lowered.

8

TRANSFORMER VAULTS

Large power transformers operate at high energy levels and can generate a considerable amount of heat. They, therefore, are treated by the NEC as not being safe to approach unless adequate safeguards are taken. One of these safety requirements involves the need to place certain transformers in a vault in order to prevent access to them by unqualified persons. The vault also serves to contain fires, insulating fluid spills, etc.

The NEC provides the guidelines for construction of the required vault in which a transformer is to be located. The structure is not just an ordinary isolated room; there are many specific rules that must be adhered to for the room to be meet requirements.

TRANSFORMERS REQUIRING A VAULT

Not all transformers are required to be installed in a vault. There are different sets of rules that apply to dry-type and liquid-filled transformers, their sizes, their location, and their operating voltage.

Dry-type transformers rated at not over 112½ kVA are not restricted in where they can be located within a building. Those over 112½ kVA located indoors must be located in a transformer room of fire-resistant construction, unless they are completely enclosed, or are separated from combustible materials by a fire-resistant, heat-insulating barrier, or by not less than 6 ft (1.83 m) horizontally and 12 ft vertically. Note that this "transformer room" is not the same as a vault. The NEC gives specific details on the construction of a vault, but not of a transformer room.

If the dry-type transformer located indoors operates at over 35kV, then it must be installed in a vault.

Dry-type transformers located outdoors have no requirement that they be installed in a transformer room or vault.

Less-flammable liquid-insulated transformers are permitted to be installed without a vault in buildings constructed of noncombustible materials and in which there are no combustible materials stored. They must,

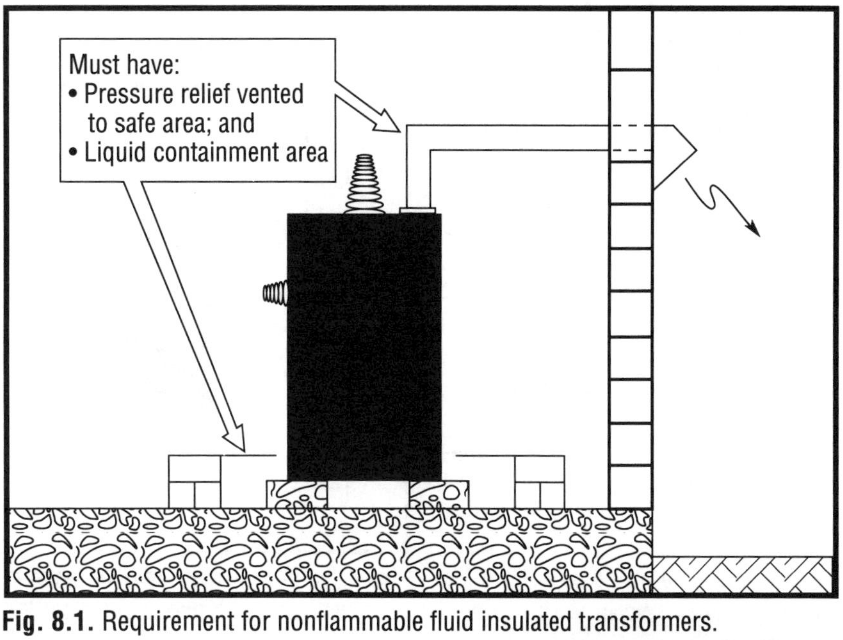

Fig. 8.1. Requirement for nonflammable fluid insulated transformers.

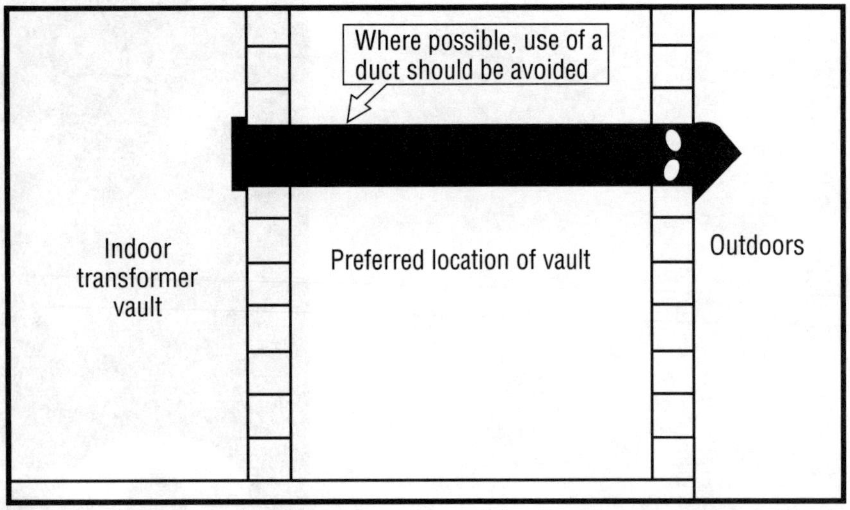

Fig. 8.2. The preferred location of a cable vault is where it communicates directly with an outside wall of the structure. When this is not possible, a duct made of fire-resistant material can be used.

however, be surrounded by a liquid-containment area. If the building does not comply with these requirements, then the transformer must either be provided with an automatic fire-extinguishing system and a liquid confinement area, or it must be located in a vault. These types of transformers operating at over 35kV must be installed in a vault.

There are no requirements that these transformers be installed in a vault if located outdoors.

Nonflammable fluid insulated transformers are those whose dielectric insulating fluid does not have a flash point or fire point, and is not flammable in air. If operating at 35kV or less they are allowed to be installed indoors with only the restrictions shown in **Fig. 8.1**. Those operating at over 35kV must be installed in a vault. There are no restrictions placed on those located outdoors.

Oil-insulated transformers, when installed indoors, must be located in a vault. There are exceptions to this rule, but they are so limited that they do not need to be discussed here. Those installed outdoors have to meet some specific safety rules, but none of these require that they be installed in a vault. Pole-mounted oil-insulated distribution transformers are covered by ANSI C2, the *National Electrical Safety Code* rather than the NEC, but since these units are mounted on poles, they obviously cannot be in vaults.

LOCATION

The general rule in the NEC that governs transformer vaults states that they must be located away from windows, doorways, combustible material, and fire escapes, or non-combustible barriers must be installed between them and the transformers. As shown in **Fig. 8.2**, however, the NEC does recognize that there are instances where it is not possible to comply with this requirement. Thus, the ending of this rule says that the rule applies, "wherever such an arrangement is practicable."

This version of the general rule is based on the 1993 edition of the NEC. If earlier or later editions of the code are applicable in the jurisdiction where the installation is located, a careful analysis must be made of any differences between the two versions.

STRUCTURAL REQUIREMENTS

Following are the specifications for the materials of construction for a transformer vault. The applicable section of the NEC states that studs and wallboard construction is not acceptable for meeting the requirements of this section. The typical requirements are shown in **Fig. 8.3**.

Walls and roofs of vaults must be constructed of materials that have adequate structural strength for the conditions. They must have a minimum fire resistance of 3 hours.

Floors of vaults in contact with the earth **(A)** must be of concrete not less than 4 inches (102 mm) thick.

Where the vault is constructed with a vacant space or other stories below it **(B)**, the floor must have adequate structural strength for the load imposed upon it by the transformer and other items, and have a minimum fire resistance of 3 hours.

Details. The NEC states that six-inch (152-mm) thick reinforced concrete is a typical 3-hr construction. An exception to the requirement for 3-hr fire ratings is allowed where the transformers are protected with automatic sprinkler, water spray, carbon dioxide, or halon; construction of a vault with a 1-hr rating is permitted.

Doorways. Vault doorways must be protected per the following set of rules. They are illustrated in **Fig. 8.4**.

Type of door. Each doorway lead-

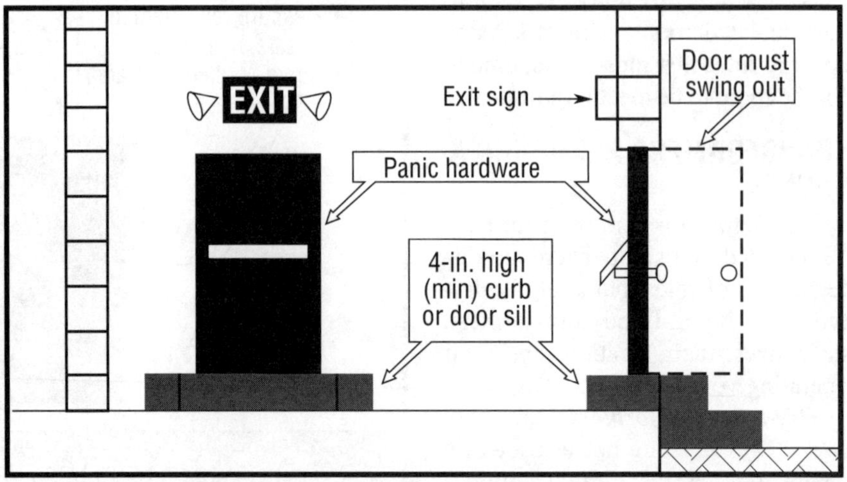

Fig. 8.4. An easy method of egress from a transformer vault is required including the use of panic hardware and outward-opening doors.

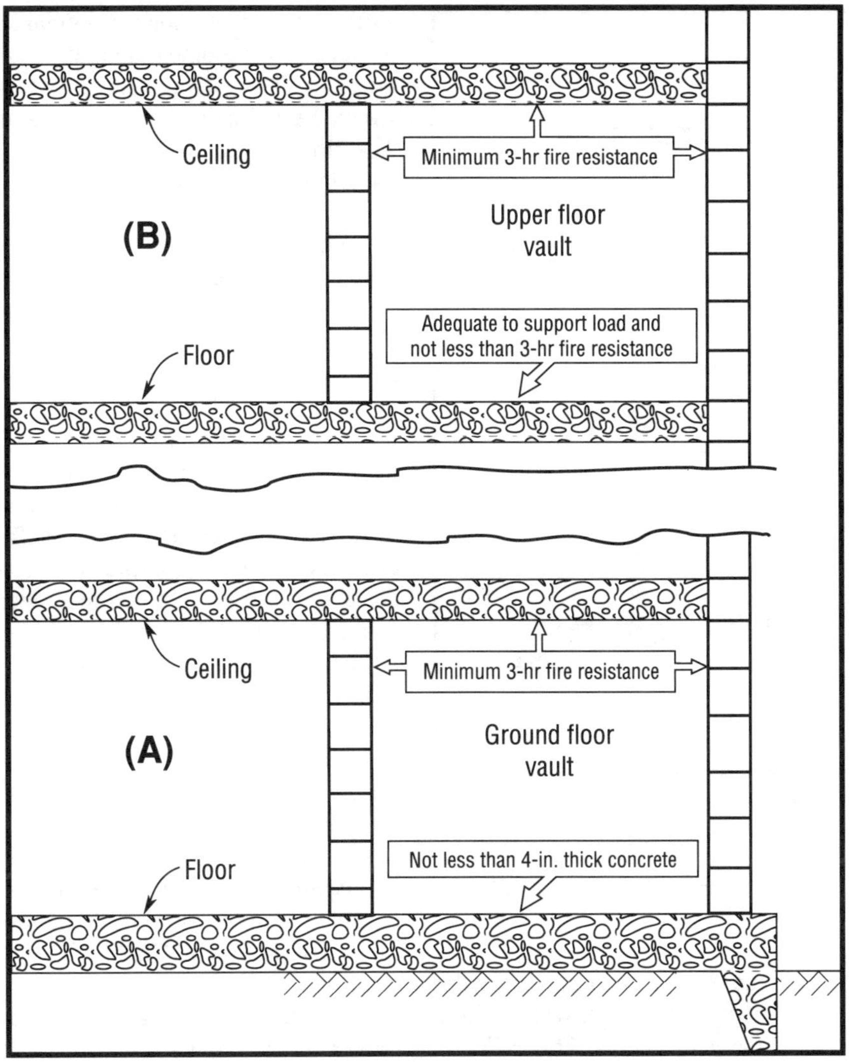

Fig. 8.3. Specifications for a transformer vaults vary according to whether it is located at ground level or in an upper level of a building.

ing into a vault from the *building interior* must be provided with a tight-fitting door having a minimum fire rating of 3 hrs. In some jurisdictions a door of similar fire rating is also required for an exterior wall opening also. This must be checked.

A 1-hr fire rated door is permitted where the transformers within the vault are protected with automatic sprinkler, water spray, carbon dioxide, or halon.

Sills. A door sill or curb of sufficient height must be provided to confine within the vault the oil from the largest transformer. In no case is the height of the sill or curb to be less than 4 inches (102 mm).

Locks. Doors of the vault must be equipped with locks for the purpose of keeping the doors locked so that only qualified personnel have access to the transformer vault.

Personnel doors must swing out from inside the vault. They must be equipped with panic bars, pressure plates, or other devices that are normally latched but open under simple pressure. The intent is to assure an easy method for persons to exit the vault even if they have lost the use of their hands due to a fire or an arc.

PROVISIONS FOR COOLING

Openings for ventilation must be provided in accordance with the following set of rules. These requirements are illustrated in **Fig. 8.5** (next page).

Location. Ventilation openings must be located as far away as possible from doors, windows, fire escapes, and combustible material.

Size. For a vault ventilated by the natural circulation of air to an outdoor area, the combined net area of all ventilating openings, after deducting the area occupied by screens, gratings, or louvers, must not be less than 3 sq in. (1936 sq mm) per kVA of transformer capacity in service. In no case can the net area be less than 1 sq ft (0.093 sq m) for any capacity under 50kVA.

Arrangement. A vault ventilated by natural circulation of air is permitted to have roughly half of the total area of openings required for ventilation in one or more openings near the floor and the remainder in one or more openings in the roof or in the sidewalls near the roof. All the area required for ventilation is also permitted to be provided by one or more openings in or near the roof.

Details. Ventilation openings must be covered with durable gratings, screens, or louvers, according to the treatment required in order to avoid unsafe conditions. All ventilation openings to the *indoors* must be provided with automatic closing fire dampers that operate in response to a vault fire. The dampers must possess a standard fire rating of not less than $1^{1}/_{2}$ hours. In addition, ventilating ducts must be constructed of fire-resistant material.

MISCELLANEOUS REQUIREMENTS

There are additional NEC rules, illustrated in **Fig. 8.6** (next page), that apply to transformer vaults. Vaults containing more than 100kVA of transformer capacity must be provided with a drain or other means that will carry off any accumulation of oil or water in the vault, unless local conditions make this impracticable. When a drain is provided, the floor must be pitched to the drain.

One of the main problems for those responsible for an electrical installation is to keep water piping and other equipment and materials out of an electric room. A vault is a specialized form of electric room, and there are several NEC rules that prohibit anything but the transformers and related equipment from entering.

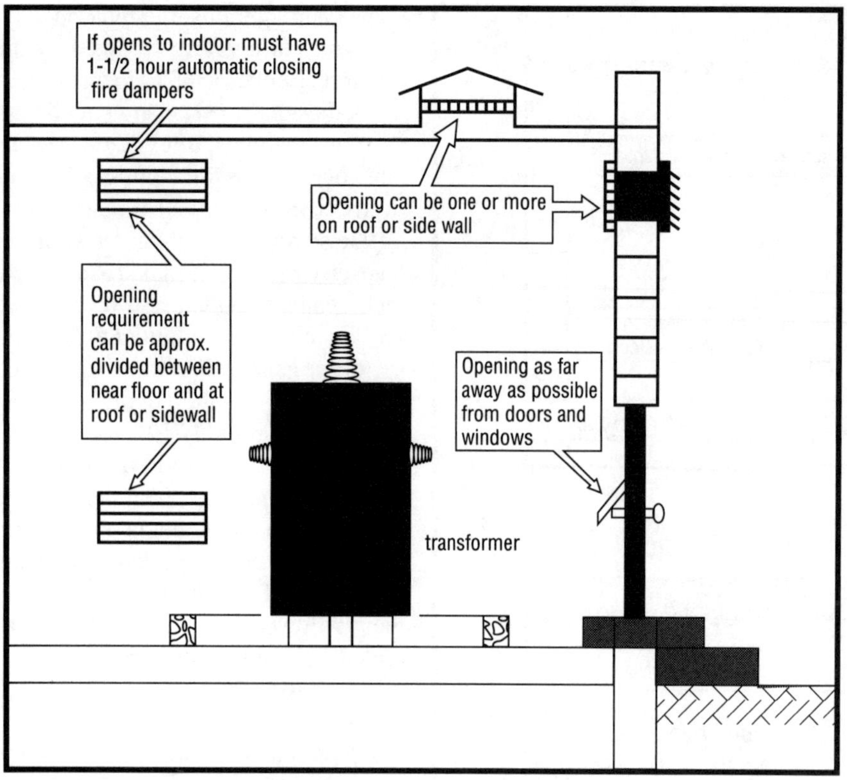

Fig. 8.5. Ventilation requirements for transformer vault.

Any pipe of duct system foreign to the electrical installation must not enter or pass through a transformer vault. This prohibition, however, is not total. Piping or other facilities provided for vault fire protection, or for transformer cooling, is not considered to be foreign to the electrical installation. Note, however, that materials must not be stored in transformer vaults.

VAULTS FOR TRANSFORMERS IN HAZARDOUS LOCATIONS

The NEC also contains rules for applying transformers that are to be installed in hazardous (classified) locations. In many cases, transformers in hazardous location are required to be installed in vaults.

Class I, Div. 1 locations. Transformers containing liquid that will burn must be installed in approved vaults. If the liquid contained will not burn, the transformer must similarly be installed in a vault unless the transformer is approved for Class I locations. Construction of the vault are as detailed previously except there are additional requirements, which are illustrated in **Fig. 8.7**. No door or other communicating opening is allowed between the vault and the Div. 1 location.

Ample ventilation must be provided for the continuous removal of flammable gases or vapors. This NEC requirement is somewhat ambiguous. The main emphasis of the ventilation requirement is to prevent a buildup of an atmosphere within the vault with a sufficient concentration of flammables to cause it to become potentially dangerous. Thus, a *positive pressure* should be maintained within the vault. The flow of air should be from the vault into the hazardous area, not vice-versa. This can be accomplished by assuring that amount of fresh clean air drawn in from the outside exceeds the amount of air exhausted from the vault.

No specific guidelines are given on the number of air changes or any of the specifics. These must be engineered to meet the needs of the installation. The only requirements that are included in the rules are that the vent ducts and openings must be of sufficient area to relieve explosion pressures within the vault, and must lead to a safe location outside the building. All portions of vent ducts within the buildings must be of reinforced concrete construction.

Class I, Div. 2 locations. Transformers in these locations must comply with the same provisions as discussed earlier for transformers in nonhazardous areas. There are no additional requirements similar to those for transformers in Class I, Div. 1 locations.

Class II, Div. 1 locations. Transformers in these types of locations are severely restricted by NEC requirements. Those containing liquids that will burn must be installed in vaults complying with the rules stated previously.

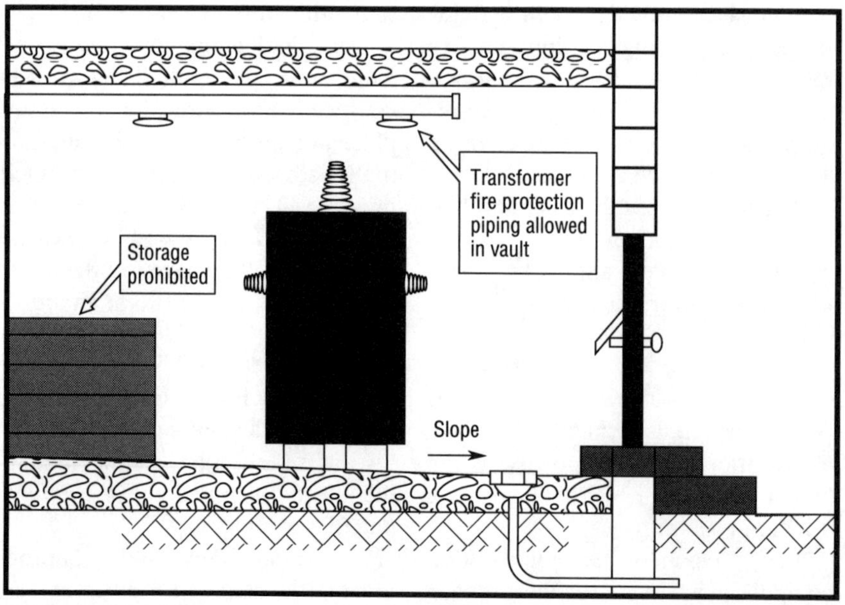

Fig. 8.6. Miscellaneous rules affecting transformer vaults.

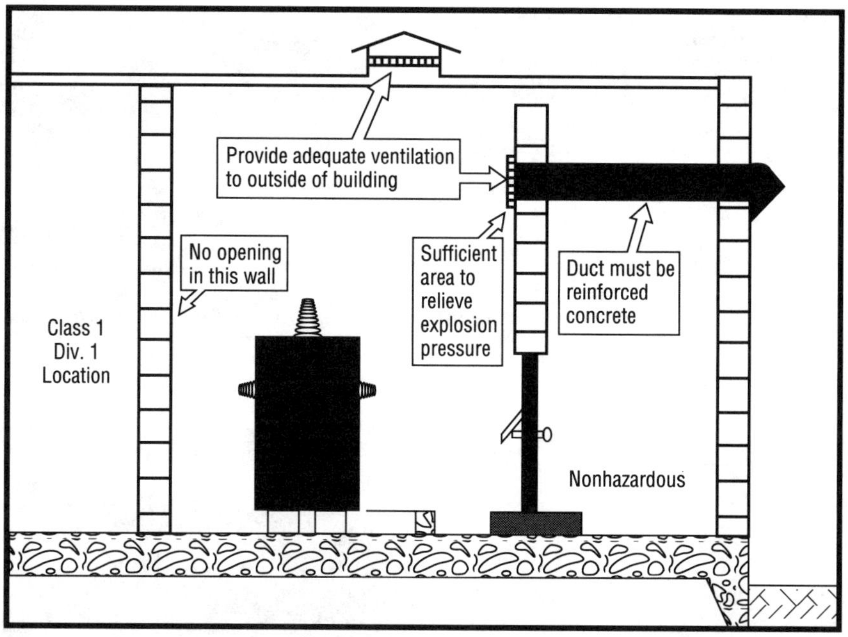

Fig. 8.7. Requirements for a transformer vault located in a building containing areas classified as Class I, Div. 1.

In addition, as shown in **Fig. 8.8**, vaults with doors opening to Class II, Div. 1 locations must have suitable seals to minimize entrance of dust; self-closing fire doors on both side of the wall; and suitable pressure-relief openings to the outside air.

Transformers containing liquids that will not burn must also be located in similar types of vaults, or else be complete assemblies (including terminal connections) approved for Class II locations.

No transformers are permitted to be installed in a location where magnesium, aluminum, aluminum bronze powders, or other metals of similarly hazardous characteristics are present.

Class II, Div. 2 locations. If the transformers contain liquids that will burn, they must also be installed in a vault meeting the requirements stated previously.

Dry-type transformers operating at not over 600V do not have to be installed within a vault as long as their windings and terminal connections are enclosed in tight metal housings that do not have ventilating or other openings.

Note that these restrictions are more severe than those imposed for transformers located in Class I, Div. 2 areas. The special allowances for less-flammable liquid insulated transformers do not apply here.

Class III, Div. 1 and Div. 2. Transformers for use in a Class III location follow the NEC rules for those installed in Class II, Div. 2 locations.

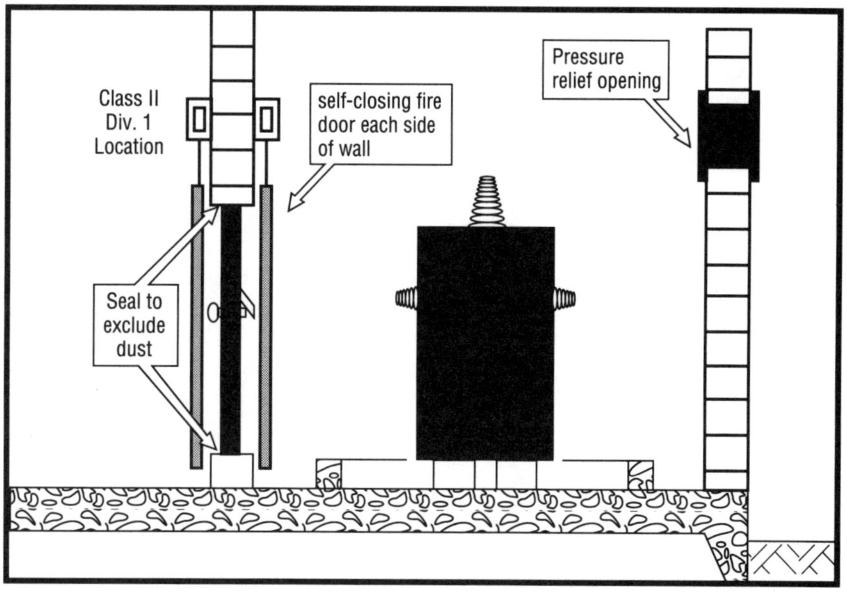

Fig. 8.8. Requirements for a transformer vault located in a building containing areas classified as Class II, Div. 1.

Transformers containing PCB are no longer being introduced into service because of the suspected effect of the fluid on the environment. Alternatives include silicone and high molecular weight hydrocarbon.

SPECIFYING LIQUID-FILLED TRANSFORMERS

9

There are may similarities in requirements between dry-type and liquid-filled transformer. The specification details for each, however, are considerably different. For that reason, specifications for liquid-filled transformers are considered in this chapter, separate from specifications for dry-types, which are covered in Chapter 10 of this book.

DRY-TYPE VS LIQUID-FILLED

The advantages and disadvantages of dry-type versus liquid-filled units depend upon the application. Dry-type transformers can usually be located closer to the load, resulting in cost savings because of shorter cable runs and lower electrical losses. A liquid-filled transformer, on the other hand, may require special construction features for the room in which it will be placed, due to fire safety reasons. This may dictate a location some distance from the load. In addition, periodic testing must be conducted on the fluid to determine its dielectric strength, water content, dissolved gases, etc.

In some applications, there is no option to the use of liquid-filled transformers. Dry-types are limited in size and voltage. On the other hand, liquid-filled types are available in almost limitless kVA and voltage ratings. Also, if requirements call for a transformer to be located outdoors, it may be less expensive to purchase a liquid-filled unit. With oil as the liquid, the cost would be lower than for a dry-type. With low-firepoint fluids, the cost would probably be comparable to a dry-type.

For liquid-filled transformers, the main cooling/insulating mediums used today are mineral oil, high-molecular-weight hydrocarbon, and silicone fluid. A high-energy arc can occur in a liquid-filled transformer, or the unit could be located in a building that catches fire. In both these cases, if there is a leak in the transformer tank, fire safety becomes an important issue. Because of hazards associated with tank rupture and the possible ignition of the dielectric, a thorough analysis covering fire safety and the possible effects on the environment should be carried out.

Some materials are covered under the Resource Conservation and Recovery Act and the Clean Water Act,

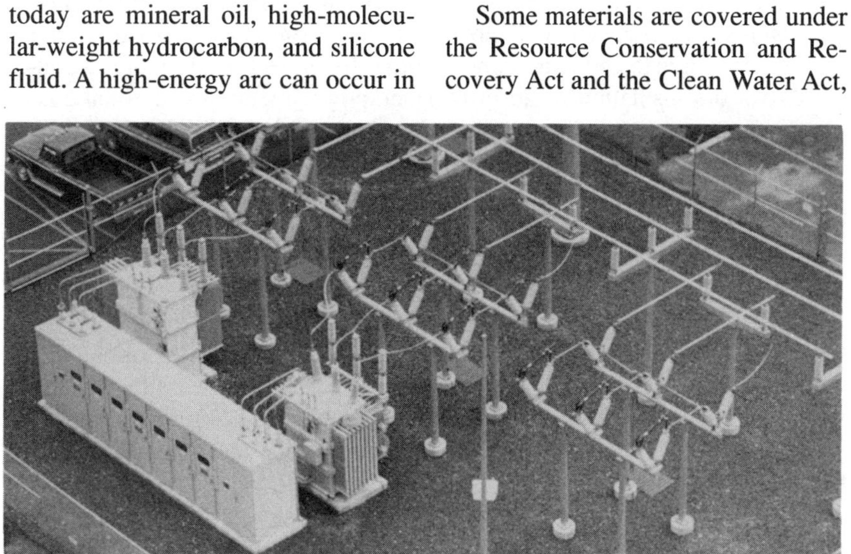

Large power transformers reduce the serving utility's transmission voltage to the utilization level that is to be distributed within the facility's power distribution system.

including requirements for special handling, spill reporting, disposal procedures and record keeping. These considerations can have an affect on installation costs, long-term operating costs and maintenance procedures.

INSULATION

Liquid-filled transformers use an insulation system of kraft or aramid paper, pressboard or aramid spacers, and a fluid that serves as both an insulating and cooling medium for the transformer.

Paper is primarily used for insulation between layers of winding material. It has an diamond-patterned adhesive backing that, when cured, solidly contains the winding.

Spacers serves as a form (which can be rectangular or cylindrical in shape) for the windings as well as a spacer between layers of the windings. The spacing is necessary to allow the insulating fluid to flow through and cool the windings and the core. Spacers are also used to insulate the windings from the core as well as to support the leads on their way out to the bushings.

Any moisture that is present in the finished core and winding assembly is purged by vacuum and oven drying processes. After removal from the oven and while still hot, all connections are retightened and the entire assembly is immersed into its liquid-filled tank. This assures that moisture will not again penetrate into the windings, and also allows the insulation to absorb the maximum amount of dielectric fluid.

The particular type of insulation to be used is rarely specified by the customer for other than large utility substation transformers or for unusual applications. Most often it is the transformer manufacturer or rebuilder's choice based upon the operating conditions the transformer must meet.

INSULATING LIQUIDS

Dielectric liquids of various types are used as an insulating medium as well as a cooling means in liquid-filled transformers. Until its use was banned because it was thought to be a potential carcinogen, PCB (askarel) was the alternative to oil for applications indoors because of its excellent fire characteristics. Alternatives have since been introduced, but none quite matches those of PCB, as seen in **Fig. 9.1**.

	Askarel	Mineral Oil	HMW Hydrocarbon	Silicone
Flash Point	195°C	150°C	284°C	300°C
Fire Point	None	165°C	312°C	343°C

Fig. 9.1. Flash and fire points of transformer liquids.

Mineral oil. A oil-filled transformer is generally the smallest, lightest, and most economical transformer that can be provided. It has excellent properties for used in transformers, but has the inherent weakness of being flammable. Its use, therefore, is restricted to outdoor installations or installed within a transformer vault if used indoors. For more on the subject of transformer vaults, see Chapter 8 of this book.

Silicone. A wide variety of synthetic polymer chemicals are referred to by the generic term "silicone." Silicone transformer liquids are actually known chemically as polydimethylsiloxane (PDMS). It is a water-clear, odorless, chemically stable, nontoxic liquid. PDMS has been in use in transformers since 1974 with a good service record.

High molecular weight hydrocarbon (HMWH) is another high-firepoint dielectric that is widely used as a transformer liquid. It has similar values for dielectric strength and constant, power factor, and thermal conductivity as mineral oil. HMWH been used as transformer dielectric coolant since 1977.

There is no established standards for testing the fire safety of transformers. Factory Mutual Research (FM) and Underwriters Laboratories (UL) both have different criteria for listing transformer liquids. Fire properties of dielectric fluids are typically classified by the following characteristics.

Flash point is the temperature at which vapors from a liquid surface will ignite in the presence of a flame.

Fire point is the temperature at the surface of a liquid that will sustain a fire.

Flame spread is defined as a series of consecutive ignitions.

Ease of ignition is defined as how readily the liquid will generate and maintain a flammable fuel/vapor mixture at the surface.

Heat release rate is the product of vaporization rate and the heat of combustion or the fluid. The higher this rate in a large-scale fire, the higher the degree of fire hazard.

Smoke and toxic gas release. Incomplete combustion caused by oxygen deficiency of a large fire in an enclosed area can produce carbon monoxide, hydrogen, and gaseous hydrocarbons, all of which present significant hazards.

Selection of the dielectric liquid depends on the transformer's application. Normally, the choice is mineral oil if the transformer is to be located outdoors. The NEC does, however have some limitations for oil-filled transformers even in those locations.

Combustible material, combustible buildings and parts of buildings, fire escapes, and door and window openings, must be safeguarded from fires originating in oil-insulated transformers installed on roofs, attached to, or adjacent to a building or combustible material. Safeguards include space separations, fire-resistant barriers, automatic water spray systems, or enclosures that confine the oil of a ruptured transformer tank. One or more of these safeguards must be applied according to the degree of hazard involved in cases where the transformer location presents a fire hazard.

Oil enclosures can be fire-resistant dikes, curbed areas or basins, or trenches filled with coarse crushed stone. Oil enclosures must be provided with trapped drains where the exposure and the quantity of oil involved are such that removal of oil is important.

Less-flammable liquid-insulated transformers, such as those containing PDMS or HMWH liquids, are per-

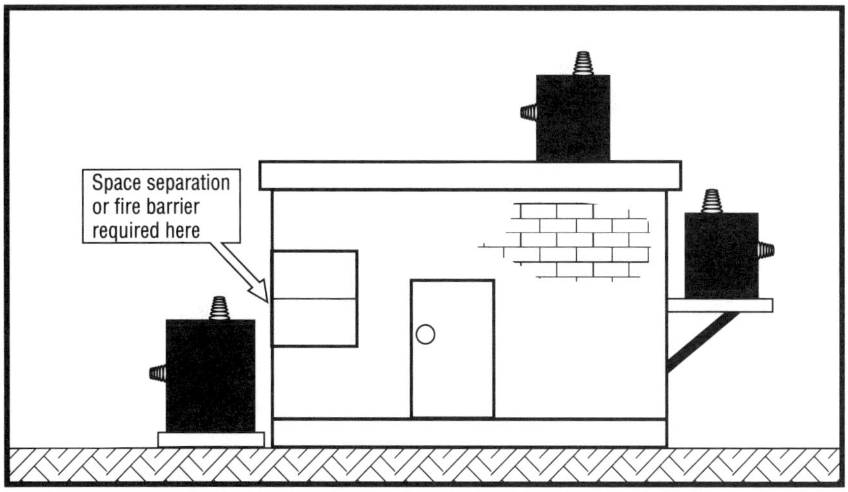

Fig. 9.2. Permitted location outdoors of transformers containing less-flammable insulating liquids.

mitted to be installed without a vault in buildings constructed of noncombustible materials and in which there are no combustible materials stored. They must, however, be surrounded by a liquid-containment area. If the building does not comply with these requirements, then the transformer must either be provided with an automatic fire-extinguishing system and a liquid confinement area, or it must be located in a vault. These types of transformers operating at over 35kV must be installed in a vault.

Fig. 9.2 shows where these transformers can be installed if located outdoors.

The selection of less-flammable liquids between PDMS and HMWH often depends upon personal preference, the liquid used in other transformers on the site, or the transformer manufacturer's recommendation.

COOLING

In cooling a liquid-filled transformer, the insulating fluid flows in the transformer through ducts and around the coil ends within a tank that contains the core and coils. Removal of the heat from the fluid takes place in external tubes. These radiators consist of headers extending from the bottom and top of the transformer tank and rows of tubes connected between the two headers. When operating within its self-cooled (OA) rating, natural convection caused by temperature differences within the tank carries the oil up through the windings, down through the cooling tubes, and back into the tank. The transformer fluid, acting as a heat-transfer medium, picks up the heat from the core and coils and dissipates it to the air via the tubes.

Auxiliary cooling fans can be provided if the transformer is to be operated above its self-cooled ratings. This is advisable when the transformer is to operate under occasional heavy overloads, high ambient temperatures, or to accommodate new loads beyond its rating. Liquid-filled transformers, because of their double heat-transfer requirement (core/coil-to-liquid and liquid-to-air), have a lower forced air rating than dry-types. In liquid-filled types the forced air (FA) rating of transformers up to 2500kVA is raised to 115% of its self-cooled kVA rating, and those of larger units to 125% of their self-cooled kVA rating.

Cooling fans can be controlled manually or automatically. Fans can be cycled on automatically based on the top oil temperature, winding temperature, or ambient temperature. Alarm contacts and remote indication are also available options.

There are some negatives associated with fan cooling.

- Losses are increased substantially. A dry-type transformer operating at 133% of its self-cooled rating will have conductor losses of nearly 1.8 times the losses at the self-cooled rating. And, there will be some losses in the form of the power required to operate the fan motors. The normal no-load losses remain constant regardless of the load.
- When additional equipment is used, such as fans, the chance of something malfunctioning increases.
- Should something interfere with fan operation so as to cause them to stop working, the result could be transformer overheating and associated problems. Thus, standards preclude the use of fans on padmount transformers.

Because of the excessive losses and

Oil-filled transformers must be separated from other structures and provisions must be made for containing any leaking oil.

Auxiliary cooling fans provide a way for a transformer to be operated beyond its normal kVA rating without exceeding the temperature rating of its insulation.

possible reduced transformer life, it is usually recommended that forced air cooling be used only to overcome temporary overload conditions. It also should be noted that load losses are less when the transformer is running closer to its rated operating temperature.

Conductor insulation can also increase the temperature rating of liquid-filled transformers. A paper insulated conductor system in a liquid-filled transformer is usually classified for 115°C operating temperature. If this rating can go up it means that the transformer's loading can be increased. Liquid-filled transformers are usually classified as 55°C or 65°C rise units. To avoid an extra initial cost, liquid-filled transformers are almost always built using paper-insulated conductors. However, when liquid-filled units are rebuilt, for a relatively small incremental increase in price, the transformer can be rewound with higher-temperature-rated conductor insulation, thus allowing an increase in rating and/or life.

LIQUID PRESERVATION SYSTEM

Temperature changes due to varying ambient temperature and load variations cause the insulating liquid to expand or contract. A completely sealed transformer tank will develop excessive positive or negative pressures under these conditions. If, however, the tank is freely vented, moisture-laden air will enter when the oil contracts. This cycling will cause the transformer's insulation system to deteriorate.

There are essentially three methods employed to overcome this problem: sealed, conservator, and positive pressure (nitrogen). Sealed tanks have a pressure-vacuum bleeder that maintains the internal pressure or vacuum within the proper operating range for the transformer. A transformer with a conservator system has some form of expansion tank or other system provided to contain the excess fluid that is forced from the tank when the liquid expands, and allows it to drain back when it contracts when cool. Large substation transformers usually are completely sealed and have inert gas (nitrogen) in the space above the liquid. Since gas is compressible, it will accommodate the normal differences in level within the tank.

The specifier will not normally be required to call out the type of system used. Alarm contacts to signal unusual pressure relief device operation are, however, useful options rather than standard.

CORE MATERIAL

The choice of metal is critical for transformer cores. Almost all transformer manufacturers today use low-loss core steel to decrease eddy currents and the effects caused by magnetic hysteresis. To achieve these objectives, high permeability, cold-rolled, grain-oriented, silicon steel is used.

It should be noted that "low-loss core material" is a relative term. While all transformers now use special core steels to achieve low losses, the steels come in several grades. Usually, grades M-6, M-4, and M-3 are used. In general, the thinner the laminates, the lower the losses. Sometimes the newer plasma etched steels having losses even lower than M-3 are used.

How low a loss should be specified? The answer depends upon the cost of the core material, the cost of losses, the transformer's loading requirements and other factors such as size limitations. In effect, an analysis should be made of the relative costs involved.

The type of core material that the transformer manufacturer will use depends upon the purchaser's specification of the losses required. Thus, it is not necessary to include a specification for the type of material with which the core is to be constructed.

TANK CONSTRUCTION AND FINISHING

Transformer tanks are customized to the transformer it is to include. External bracing must be accurately sized in order to assure a rigid structure that is capable of withstanding the stresses imposed on it. Welding must be first class to avoid even pinhole opening that would allow moisture into the tank. The tank is thoroughly leak tested following

A wound core made of amorphous metal for a distribution transformer greatly reduces core losses. The atomic structure of the metal, like glass, is random rather than in an orderly repeated array. This and the thickness of metal accounts for its very low hysteresis and eddy current losses.

fabrication.

The cover and all openings for bushings, gages, and any other penetration get special attention. They are fitted with dimensionally stable gaskets to prevent escape of the insulating liquid or entrance of moisture. These gaskets have to last the lifetime of the transformer.

Before painting, internal and external tank surfaces are blasted clean and protected against corrosion by a coat of primer, followed by coats of enamel that are baked on. A finish coat of enamel is then sprayed on and air dryed.

The standard paint color for transformers is usually light gray ANSI #70 (Munsell No. 5.0BG 7.0/0.4). If another color is required, this would have to be included in the specification.

An option that is sometimes called for is the painting of the interior of the tank. When this requirement is specified, white is usually called for. Its purpose is to allow better visual inspection of the internals, and increasing the likelihood of locating any object that may have settled on the tank floor.

STANDARD RATINGS

Liquid-filled power transformers come in so may sizes and configurations that it is difficult to generalize about their ratings. They can, however, basically be divided between small distribution types that are located on poles, structures, or are padmounted; and larger substation types that are mounted either outdoors or indoors. The substation types are further subdivided into stand-alone units or loadcenter types that are to be integrated into a unit substation.

Distribution transformers. These are small single-phase units ranging from about 25kVA to 100kVA, or 3-phase units ranging from about 30kVA to 300kVA. They are used mostly to serve individual loads, such as homes, light commercial, and similar installations. They also are used to step down the distribution voltage to that used in a facility. Primary voltage is usually in the 600V to 5kV range, with a secondary voltage 120/240V, 208Y/120V, 480V, 480Y/277V, and others.

Small padmounted liquid-filled units such as this are provided with incoming and outgoing cable compartments for ease of connection.

Substation transformers. Generally, these units range from 500kVA up to 30MVA or larger. In utility service, the larger units are either use to step up the voltage for transmission, or to step down the voltage to meet the users need. Within a facility, the smaller units are used to step the primary distribution voltage down to the utilization level. The primary voltage here is principally in the 5kV to 69kV range. Secondary voltage is mainly in the range of 480V to 13.5kV. Larger units in utility service cover high- and extra-high voltages.

Loadcenter transformers. These are 3-phase transformers in the 225 to 5000kVA range that are part of a package usually consisting of a medium-voltage primary disconnect device, the transformer, and low- or medium-voltage switchgear, switchboard, or similar equipment. The primary and secondary bushings extend out from the sides of the tank and are bussed to the other components. The bushings are not visible from outside the metalclad lineup of equipment. Voltages normally used can be anywhere from 2.4 to 35kV for the primary, down to a 208Y/120V secondary.

Special types. There are many transformers that are designed to handle the requirements of a specific application. Among these are those to provide traction power, rectifiers, pulsed loads, etc.

Selection here depends entirely on the engineering and design of the power distribution system of which the transformer is to be part of.

BIL

The basic impulse insulation level (BIL) is a measure of the ability of the transformer's insulation system to withstand very high voltage surges. Standards have been devised setting the BIL level for the specific voltage level at which the transformer is to operate, and for the type of tests that all transformer manufacturers must make on their equipment to demonstrate the BIL of their equipment.

An impulse test wave that increases to its crest value in 1.2 microseconds, and decreases to one-half of the crest value in 50 microseconds (a 4 x 50μs voltage wave) is set as the standard upon which the test is based. This waveshape is reasonably similar to what could be expected to be generated by lightning.

As an example, a 7200V transformer operating on a 15kV class distribution system has a standard BIL of 95kV. This means that the 15kV class transformer's insulation must be able to withstand the standard voltage wave having a crest value of 95kV. Transformers used on 34.5kV systems have a BIL of 125, 150kV, or even 200kV.

Fig. 9.3 (next page) illustrates some typical BIL levels and the associated impulse test requirements. Note that there are two test values; one in which the insulation is subjected to the full wave, and the other in which the wave is chopped at about 3 microseconds. The crest value of the chopped wave is 115% of the full wave value. The significance of these values is best analyzed by referring to ANSI Standard C57.12.01.

IMPEDANCE

The combination of resistance (R)

Nominal System Voltage	BIL	Insulation Class	Low Frequency Test	Impulse Tests		
				Chopped Wave		Full Wave
kV	kV	kV	kV	kV Crest	Min. Time to Flashover (Microseconds)	kV Crest
1.2	30	1.2	10	36	1.0	30
2.4	45	2.5	15	54	1.25	45
4.8	60	5	19	69	1.5	60
8.32	75	8.7	26	88	1.6	75
14.4	95	15	34	110	1.8	95

Fig. 9.3. Typical dielectric tests of winding insulation.

and reactance (X) that is part of a transformer is expressed in terms of impedance (Z). Mathematically, impedance is defined by the equation:

$$Z = R + jX = R + j(2\pi fL) \text{ ohms},$$

for an inductive circuit.

The impedance of a transformer determines how it reacts when installed as part of a power distribution system. It is particularly important in limiting the amount of current that will flow in the transformer secondary during short-circuit conditions. The higher the impedance, the lower will be the current.

Specifying the amount of impedance required is tricky since there are many considerations that must be taken into account. For instance, the required withstand ratings of downstream protective devices must be factored into the decision. Yet, circuits feeding loads with significant amounts of nonlinear characteristics would be better served by a transformer having a lower impedance.

Transformers are usually bought based on standard impedances, such as those shown in **Fig. 9.4**. Basically, these impedances vary according to the kVA rating of the unit. Special ratings, such as 8% rather than 5.75%, are also available. Only a thorough engineering evaluation of distribution system needs can determine what level will best be suited.

kVA	Percent Impedance
112.5	2.0
150	2.0
225	2.0
300-500	4.5
750-3000	5.75

Fig. 9.4. Typical minimum standard impedances. Transformers over 1000kVA are available with 8% impedance.

TAP CHANGERS

One of the main functions of a transformer is to maintain a steady optimum voltage level to its load. This is not always easy to achieve in situations where long runs of cable are involved, or where the load varies significantly over time. Since transformer secondary voltages represent standard values, if a heavy load is at a considerable distance from the transformer, the voltage drop in the cable may reduce the voltage at the load to a less than desirable level. Similarly, if a very heavy load is cycled on or off periodically, the system voltage level may be significantly affected.

In order to compensate for these situations, transformers are usually provided with taps on their windings. These are generally two $2^1/_2$% taps above the nominal voltage rating of the transformer, and two $2^1/_2$% taps below the rating. In this manner, if the voltage is too low, it raised $2^1/_2$ or 5%, or lowered a corresponding amount if the voltage is too high.

These taps are not readily available in liquid-filled transformers. Thus, some means must provided to make these voltage adjustments externally. Tap changers accomplish this purpose. There are two basic types of tap changers.

Deenergized tap changers are used when it is likely that the voltage in the system will have to be adjusted only very occasionally. In these cases, it would be possible to schedule a shutdown of the system while the tap position is changed. Still, it would not be convenient to have to drain the transformer liquid in order to physically disconnect and reconnect the taps. Thus, provisions are made for the tap changes to be accomplished via gang-operated contacts within the tank. These are moved from tap to tap via a handle that protrudes through the tank wall. Provisions are made for the handle to be padlocked in place. Good safety procedures must be in force at the location to assure that the tap changes are not made under load.

Load tap changers offer the possibility to make the tap changes while the system is energized and under load. This requires shunting the load current from one winding tap to the next via a motor-operated mechanism. This is a two step operation where first one switch leg transfers to the adjacent tap (in the direction to either increase or decrease voltage), and then the second switch leg transfers to the same tap as shown in **Fig. 9.5**. Some means is provided (a reactor or autotransformer) to limit the resulting circulating current to a safe value during the interval when two adjacent taps are bridged. Because of the circulating and load current, arcing always takes place as the power circuit is transferred form tap to tap. Key to solving the problem is the breaking of the arcs within vacuum interrupters.

Very precise control systems are used to regulate the motor that drives the tap-changer switches. Some sort of voltage sensing circuitry is usually involved. The control panel is mounted on the transformer and selection and indication of tap location is displayed.

Automatic tap changers are available for extreme cases where precise voltage must be maintained while on a system in which the voltage fluctuates frequently. The sensing and control mechanism is very sophisticated and, thus, is costly.

The specifier normally only needs to specify whether a deenergized or load tap changer scheme is preferred. As an option, it is possible for a window to be installed that will allow observing tap-changer position.

BUSHINGS

Connections to the medium- or

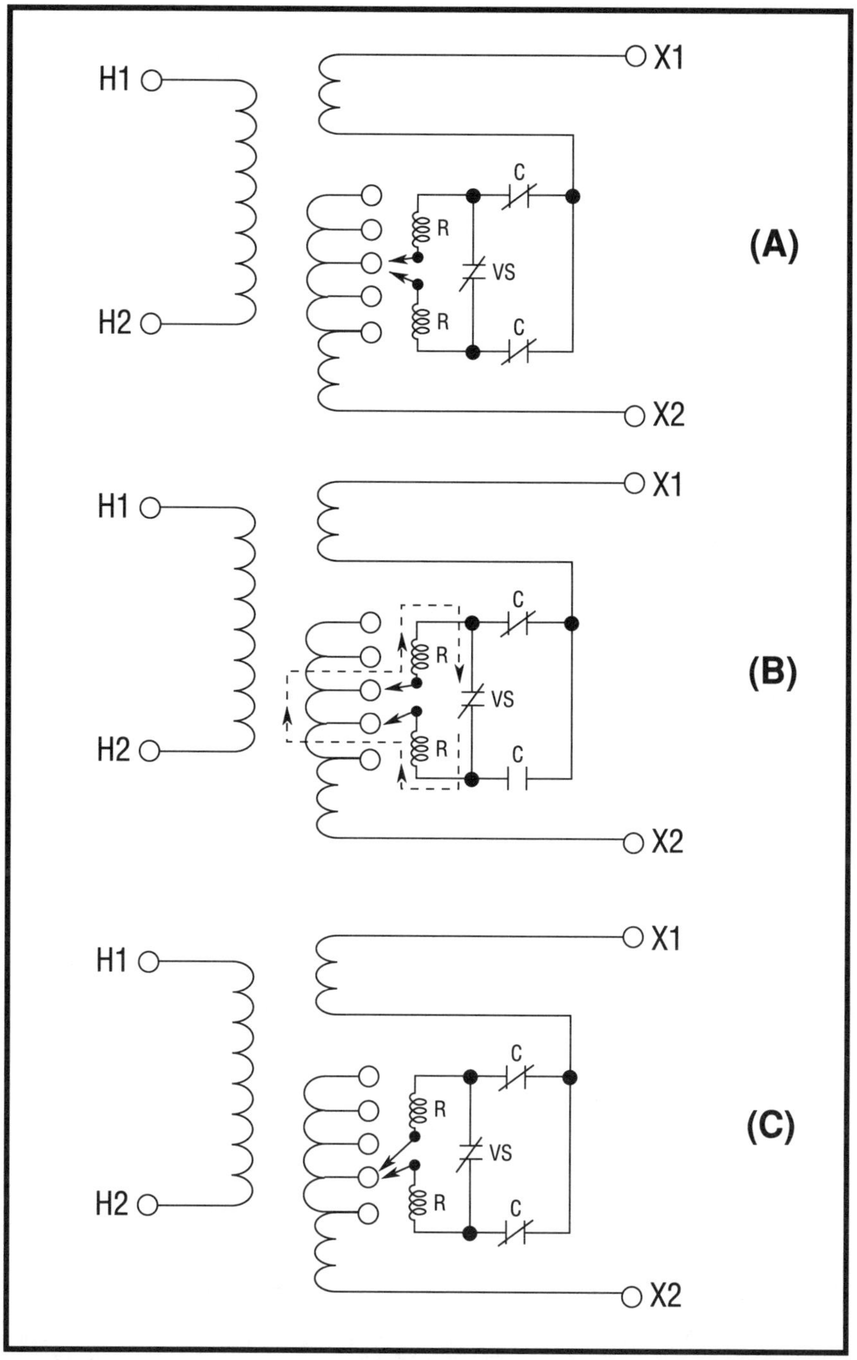

Fig. 9.5. Schematic diagram of a load tap changer; R = reactor, VS = vacuum switch, C = bypass contacts. In **(A)**, the rated-voltage center tap is being used. In the first step in changing the tap **(B)**, one switch is transferred to the new position, setting up a circulating current (dotted line). The second contact is then transferred **(C)**. An actual load tap changer can have 16 taps above and 16 below rated voltage.

Bushings can be located on both top and side such as this transformer that is to be connected to a lineup of LV switchgear.

high-voltage windings and to the lower-voltage secondary windings of a liquid-filled power transformer is through bushings located either on the top and/or side of the tank. High-voltage bushings are often oil insulated, consisting of a hollow copper tube covered with insulation, an exterior housing of porcelain, an annular space between the two for the oil, an oil expansion chamber, and a mounting flange and ground sleeve. A metal foil layer is included near the outer insulation surface with a lead running to an external connection. This is used for power factor testing of the bushing.

The standard terminal arrangement for cover bushings on 3-phase transformers is shown in **Fig. 9.6**. The bushing ratings and external creepage distance is listed in **Fig. 9.7** (next page). Depending upon the atmosphere in which the transformer is to be located, the need for bushings with extra creepage distance can be specified.

Bushing type current transformers (CTs) can also be supplied. These are relay types, but can in some cases have sufficient accuracy to be acceptable for metering as well as relaying. This option is particularly important for larger, critical transformers where differential protection is desired. The CTs monitors the current in the higher-voltage and lower-voltage circuits and instantly detects any unbalance caused

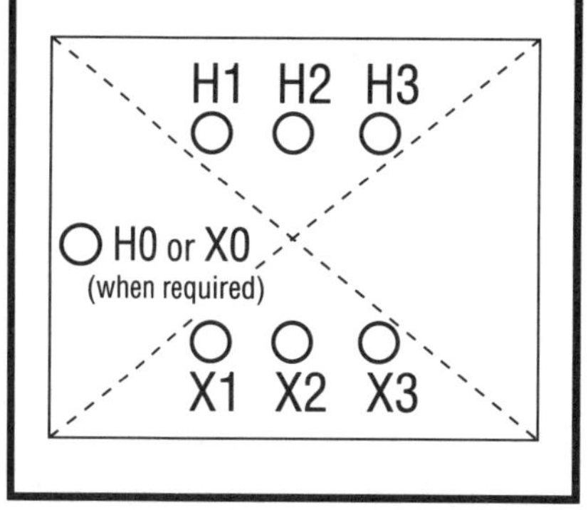

Fig. 9.6. Standard terminal arrangement for cover bushings on a liquid-filled 3-phase power transformer.

Bushing BIL-kV	Bushing Voltage Rating (kV)	Minimum Creepage Distance (inches)	Extra-Creepage Bushings Distance (inches)
110	15	11	Use Next Higher Rated Bushing
150	25	17	
200	34.5	26	
250	46	35	
350	69	48	66
550	115	79	92
650	138	92	114

Fig. 9.7. Typical bushing ratings and external creepage distance.

by a beginning failure. A relay immediately signals a primary circuit breaker to remove the transformer from the system even before the conventional overcurrent relays can operate.

SURGE ARRESTERS

Voltage surges may appear on a utility's transmission lines or on the facility's distribution system. The reason can be a direct or nearby lightning stroke, or switching of the power system. This can cause damage to transformer windings. Surge arresters can be specified that will provide a temporary low-impedance path to ground so that the voltage surge will be directed harmlessly to the earth. Once past, the insulation value between the circuit and ground is reestablished, preventing the system voltage from producing a flow of follow-through current.

Surge arresters are selected for application to the electrical system based on system characteristics, and then transformer BIL is specified to allow sufficient margin of protection between surge arrester operating characteristics and transformer insulation strength.

SELF-PROTECTED UNITS

Smaller pole-mounted single-phase and 3-phase distribution transformers can be supplied as self-protected units with an internally mounted protective link in series with the high-voltage windings to disconnect the transformer from the line in the event of an internal coil failure. In addition, an internally mounted, thermally controlled secondary circuit breaker is supplied for overload and short-circuit protection.

As an option, the circuit breaker can be connected to an external signal light when a predetermined winding temperature has been reached. This serves as a warning before tripping. If the signal is unheeded and the breaker trips, the breaker can be reset and the load restored by an external handle. If the circuit breaker will not hold in, this is an indication of a long-term overload condition. In an emergency, the breaker trip temperature can be reset externally and the breaker closed until a larger unit is available to replace the transformer.

Lightning arrester(s) are integrally mounted on the outside of the tank for overvoltage protection of these units.

NEUTRAL GROUNDING RESISTORS

Power system neutrals are sometimes grounded through a resistor. These are often coordinated and specified with the transformers.

In resistance grounding, the neutral is connected to ground through one or more resistance banks. It acts to limit the amount of ground-fault current that can flow by adding resistance to the possible ground circuits from the fault back to the neutral of the system. This limits the magnitude of the current to as small a value as possible while allowing the operation of protective relaying.

The reasons for limiting ground-fault current are:

- reduced current results in less burning and melting in the faulted electrical equipment;
- there is reduced mechanical stress in the circuits and apparatus because of the lower fault current;
- the resistance gives ground-fault current a direct path to the neutral, reducing damage to other equipment caused by stray ground-fault currents; and
- voltage drop that occurs during the fault and at clearing is reduced.

Application requirements for neutral ground resistors vary widely ac-

Pole-mounted distribution transformers can be specified to be self protected. Overload and short-circuit devices are located within the tank, and surge arresters mounted on the outside.

Neutral grounding resistor is shown before the safety screening was added to the enclosure. The stainless steel resistors are ceramic-core ribbon-wound units arranged in tiers within a steel frame. Each unit is insulated from the tier structure and this, in turn, is insulated from the frame. The frame assembly is mounted on insulators for systems above 600V to provide isolation from ground.

cording to duty. IEEE standards have established their thermal time ratings.

10-second-rated resistor limits the ground-fault current for purposes of activating protective relays and reducing equipment damage to a minimum. It depends upon a ground-fault relay's operation to prevent equipment damage.

1-minute-rated resistor serves to limit ground-fault currents on outgoing feeders during faults to ground. They reduce damage at the point of a fault by limiting the ground-fault current and also limiting overvoltages between good phases and ground. Maximum ground-fault current in such cases is often less than the full-load current of the feeder. Use of a device with a 10-second rating for this service is not satisfactory because it may be required to withstand several faults in rapid succession.

Extended-time rated resistor is used where the ground fault is permitted to persist for longer than 10 minutes. Resistors with this rating are used to permit fault current to flow until appropriate actions can be taken to eliminate the problem. The permitted temperature rise of these units is higher than standard, assuming that the operation at the maximum rise will not be required for more than an average of 90 days per year.

Continuous-rated resistor is applied in power substations and generating stations where it is not permitted to shut down the source of power to eliminate faulted conditions. A neutral grounding resistor is used to limit fault current to an acceptable level.

Neutral grounding resistors should be specified with provisions to prevent personnel from coming in contact with either the units or their support frame. For indoor locations, this protection can be provided by specifying resistors surrounded at top and sides by a freestanding enclosure that provides adequate electrical clearance. Entrance of the incoming neutral connection is made through the top side with an entrance bushing.

In outdoor installations, the resistors can be specified to be on mounting stands so that the hot end of the entrance bushing is far enough off the ground to meet the appropriate electrical codes and also provide a direct connection to the power transformer neutral. A screened enclosure is normally provided around these resistors.

Current or potential transformers for indication of ground-fault current is quite common. The turns ratio specified is dependent upon the current or voltage rating needed to activate the relaying or protective devices. Current transformers are often mounted within the neutral grounding resistor screened enclosure.

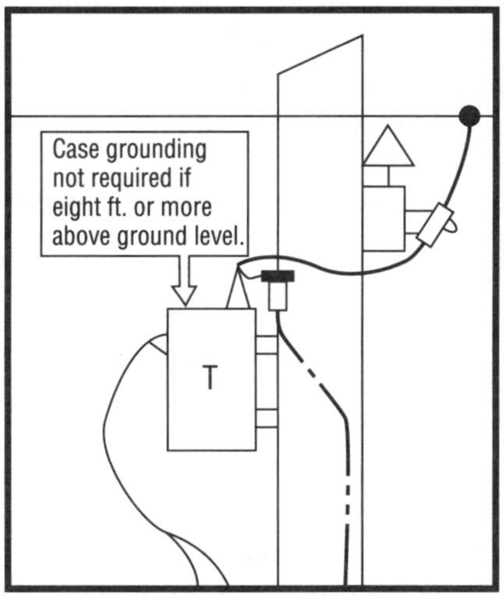

Fig. 9.8. An exception to the need for grounding the tank of a transformer.

ACCESSORIES

Grounding. NEC Article 250 requires that all noncurrent-carrying metal parts of electrical equipment be grounded. This includes the metallic casing of transformers. An exception is made for transformers located 8 feet or more above grade level, as shown in **Fig. 9.8**. Provisions for grounding the transformer tank are standard. Large transformers are supplied with grounding pads on diagonally opposite corners of the tank base for grounding connections.

Nameplate. Various standards, including the NEC require that a transformer have a nameplate so that it can be applied and serviced properly. Those provided are usually stainless steel and contain a diagram of the power circuit, kVA rating, voltage rating, operating temperature, impedance, and other characteristics. The NEC also requires that the amount and type of insulating liquid be specified.

Winding temperature. While it is not a standard accessory, if specified provisions can be made to monitor the winding hot-spot temperature. An indicator/relay is mounted on the outside of the tank. Its contacts can be used to turn cooling fans on and for remote alarms.

Liquid temperature. This is a standard feature that consists of a thermometer and external dial gage that indicates the temperature of the oil at the top of the transformer (likely to be the warmest).

Liquid level. This is another standard accessory for liquid-filled transformers. It indicates changes in liquid level, with a range above and below the level when the transformer is operating at 25°C.

Pressure/vacuum. As the dielectric liquid expands and contracts, the pressure within a sealed tank liquid

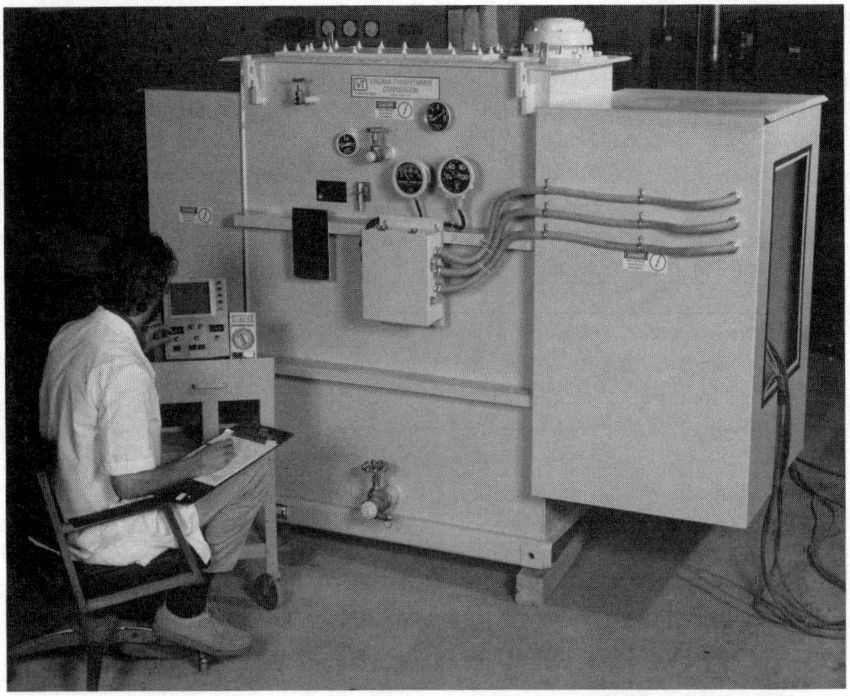

Testing upon completion is very important for large liquid-filled transformers.

preservation system will rise and fall, even dropping below atmospheric. The gage provides a way of monitoring the system.

Pressure/vacuum bleeder. This standard accessory maintains the internal pressure within operating limits.

Pressure relief. Sudden drastic rises in pressure, which could be caused by and internal fault, is relieved by a mechanically operated, self-reclosing device. While it is a standard accessory, optional alarm contacts can be specified if it is desired to remotely signal the operation of the device.

Fault pressure relay. This is an alternative way of monitoring the presence of a fault. It reacts to the primary shock wave transmitted through the insulating liquid. This is not a standard device and must be specified if required.

Surge arresters. Protection against external voltage surges can be provided on the transformer tank if specified. They can be station, intermediate, or distribution types, depending on the application.

Upper and lower filter valves and drain. These standard items are needed to check from time to time on the quality of dielectric liquid, and to filter it when needed.

Gas sampling valve. This is a standard item that vents gasses from the tank of large power transformers.

TESTING

Power transformers are tested once they have been completed to assure that they meet specifications. Some of these are standard tests conducted by the manufacturer, while others are done only if specified by the client. All are done in accordance with ANSI standard test codes.

Standard tests. The following items are checked and routinely included in test reports available to the client.

- Resistance measurements for all windings on the rated voltage connection of each unit and at the tap extremes of one unit only of a given rating.
- Ratio tests on the rated voltage connection and on all tap connections.
- Polarity and phase-relation tests on the rated voltage connection.
- No-load loss at rated voltage on the rated voltage connection.
- Exciting current at rated voltage on the rated voltage connection.
- Impedance and load loss at rated current on the rated voltage connection of each unit and on the tap extremes of one unit only of a given rating.
- Applied potential tests.
- Induced potential tests.
- Winding temperature rise for standard transformers determined from basic design data that has been verified by test results of similar transformers.

Optional tests. If specified, the following tests can be made and results made available.

- Other temperature tests where the client specifies the exact ratings and number of units on which heat runs should be made.
- Impulse test to confirm that the transformer is able to withstand the voltage surges that will occur in service. The ANSI standard impulse test consists of one reduced full wave; two chopped waves; and one full wave.
- Audible sound test to determine the audible sound level.
- Corona test to determine if the transformer's insulating system is being overstressed. Corona scars, tracks, can physically erode insulation, can cause chemical decomposition that degrades the insulation, and can decompose the dielectric liquid. Thus, it is important to verify the integrity of the insulation system before the transformer is shipped. Passing applied and induced voltage tests does not prove that the transformer is free of corona.

SPECIFYING DRY-TYPE TRANSFORMERS

10

Time spent in specifying a dry-type transformer often is in direct proportion to the size of the unit. Small lighting transformers get just a cursory look at the connected loads and a decision is made to choose one with the next larger kVA rating relative to the anticipated load. Conversely, large transformers of this type with ratings over 500kVA are closely evaluated because they represent larger investments.

This chapter will review some of the most important items that should be kept in mind when preparing a specification for a dry-type transformer. A wide range of products could be included under this name, including small single-phase control transformers. The discussion here, however, will only cover items that are more likely to require specification by the purchaser of larger units.

APPLICATION OF THE UNIT

Sizing a transformer for a particular application with regard to its life expectancy requires a good understanding of the insulation characteristics and the temperature loading, which requires a careful analysis of the load profile (both amplitude and duration) the proposed unit will experience. In addition, there are some standard parameters that, if the intended normal operating conditions will exceed, must be accounted for in the specifications. They include:

- input voltage and frequency at nominal values;
- input voltage and load current are approximately sinusoidal;
- unit installed at an altitude of less than 1000 m (3300 ft);
- no damaging fumes, dusts, or vapors in the operating environment;
- ambient temperature does not exceed 30°C as a daily average, or 40°C at any time, and does not fall below -20°C; and
- kVA ratings can be maintained without overloads.

If the application is such that some of the above conditions cannot be met, it then becomes important to work closely with the manufacturer in developing the transformer's operating characteristics and/or size to compensate for the particular situation. For example, should the ambient temperature exceed standard conditions or should the unit be installed at a high elevation, etc., then an appropriate solution might be a larger transformer with a loading less than the unit's kVA rating (in effect, derating the transformer).

Derating the transformer may also be required if the load is expected to consist of a significant amount of nonlinear loads. This subject is covered at length in Chapter 7 in this book.

DRY-TYPE VS LIQUID-FILLED

The advantages and disadvantages of dry-type versus liquid-filled units depend upon the application. Dry-type transformers can usually be located closer to the load, resulting in cost savings because of shorter cable runs and lower electrical losses. A liquid-filled transformer, on the other hand, may require special construction features for the room in which it will be placed, due to fire safety reasons. This may dictate a location some distance from the load.

Dry-type transformers are available for either indoor or outdoor installation. They can operate indoors under normal ambient conditions and are best suited for duty where load demands are not severe.

For outdoor operations, dry-type transformer enclosures will usually have a NEMA 3R enclosure with louvers for ventilation. However, these transformers can still be adversely affected by hostile environments (dirt, moisture, corrosive fumes, etc.) because their windings are exposed to the air. In this type of environment, a dry-type transformer with sealed (encapsulated) construction providing winding protection is recommended. Even these transformers are subject to attack in certain environments, and dust and dirt must be prevented from blocking the cooling passages through the transformer windings.

DRY-TYPE INSULATION SYSTEMS

One of the most important items that must be specified is the type of insulation system. Dry-type transformer insulations provide dielectric strength while enabling the transformer to operate within certain thermal limits. Dry-type transformers are available with several different classes of insulation systems:

- Class 105;
- Class 150;
- Class 185;
- Class 220; and recently
- Class 250.

At one time, letter designations were used for transformer ratings. **Fig. 10.1** shows the correlation between the old system and the new designations. What was previously termed a Class H transformer is now known as a 150°C rise transformer with a Class 220 insulation system.

The insulation ratings represent the sum of the maximum ambient temperature (40°C is standard), plus the average winding temperature rise (the transformer rating), plus the differential between the average winding temperature rise and the highest temperature of the winding. An accepted maximum 30°C above average coil winding temperature is used as the hot-spot allowance (except for Class 105, in which it is 10°C). For example, Class 220 insulation is derived from the 40°C maximum ambient, plus the 150°C average winding temperature rise, plus the 30°C maximum hot-spot.

Almost all U.S. transformer manufacturers use Class 220 insulation. By specifying a lower temperature rise unit, additional thermal allowance is provided. Temperature rise ratings of transformers can be altered by increasing the conductor size of the winding, which reduces the resistance and resultant heating; and by derating a larger higher-temperature-rise transformer.

Old terminology		New terminology			
Insulation rating	Transformer rating	Insulation rating	Transformer rating	Ambient temperature	Hot spot allowance
A	A	Class 105	55°C rise	40°C	10°C
B	B	Class 150	80°C rise	40°C	30°C
F	F	Class 185	115°C rise	40°C	30°C
H	H	Class 220	150°C rise	40°C	30°C

Fig. 10.1. Old and new transformer and insulation system ratings.

LIFE EXPECTANCY AND TEMPERATURE

The nameplate kVA rating of a transformer represents the kVA that will result in the rated average winding temperature rise when the unit is operated at 100% of rated kVA under normal service conditions. When operating under these conditions, the result should be a "normal" life expectancy for the transformer.

A general rule-of-thumb says that a 30-year life can be expected for a transformer with a 220°C insulating system that has a winding hot-spot temperature allowance of 30°C. Operating such a transformer at rated kVA on a continuous basis with a 30°C, 24-hr average ambient (40°C maximum ambient) should equate to a normal useful life.

It should be recognized that the life expectancy of transformers operating at varying temperatures is not accurately known. Fluctuating load conditions and changes in ambient temperatures make it difficult, if not impossible, to arrive at such definitive information. However, if a transformer is operated under normal conditions, it could easily last longer. A 40-year life span is not unusual and some transformers have exceeded that.

CORE AND COILS

Cores are built in one of two ways: wound cores, or stacked laminations. The laminations consist of high-quality grain-oriented silicon steel, stacked in either butt-and-lap or miter-cut construction, as noted previously in Chapter 2. The stacked cores are configured as a rectangular core or a cruciform core. The choice between wound or stacked cores, but-and-lap vs miter-cut used for a particular transformer depends upon how to achieve the best combination of efficiency, size, and weight. The choice is normally made by the manufacturer based on meeting the performance characteristics required by the specification. Thus, these considerations do not have to be addressed when writing a specification for a transformer.

Larger kVA units and higher voltage class transformers usually employ cruciform cores. The stepped shape allows the closest possible coupling with round coils used in the larger units. This configuration best supports the coil and resists the crushing effect during short-circuit conditions. On the other hand, a rectangular core and coil provide the best use of space.

Coils can be either rectangular, or round, and wound in several different ways. Except in unusual cases, the specific configuration is left to the manufacturer's discretion in meeting the equipment specification.

Barrel wound coils are generally used for windings of dry-type transformers where layer-to-layer and turn-

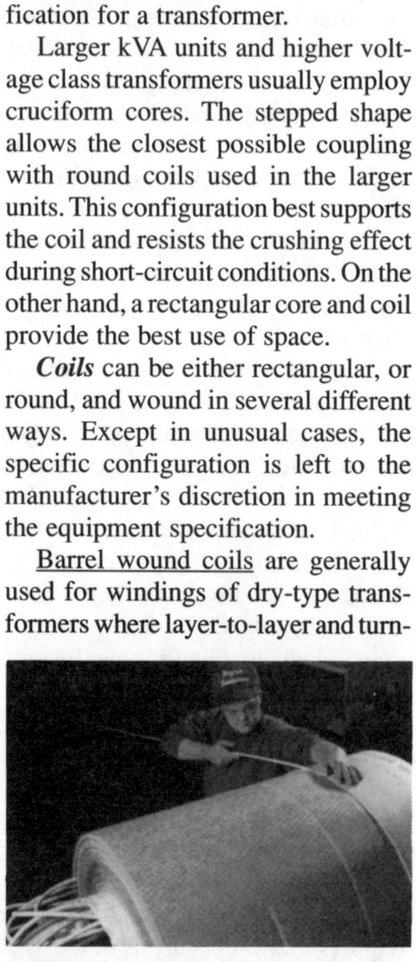

Applying a conductor of a barrel-wound winding.

to-turn voltage stress is low, but high current carrying capability is needed. Paper insulated wire is wound from one end of the winding to the other, then covered with solid-sheet insulation. Spacers are then strapped in place around the periphery and the next layer of coils wound over them. This is repeated until the proper number of coils are in place. The gap between windings allows for penetration of varnish to the finished assembly. The space between the windings also serve as air circulation channels of the completed transformer.

Applying a conductor of a random-wound winding.

Random wound coils are used where higher voltages and relatively low currents are involved. An insulated conductor is wound into slotted spacers arranged around the circumference of the coil. The coils are connected in series to provide a high capacitance to distribute impulses through the winding.

Continuous disc wound coils are used for units having both high voltage stress and high current. Multiple sections are formed with a rectangular cross-section conductor wound turn-to-turn. Individual spacers arranged radially provide support to each section. The continuous connection of all sections in series provides excellent impulse distribution.

Applying a conductor of a disc-wound winding.

Sheet windings are often used in dry-type transformer windings rated 1000V or less. In effect, it is a barrel-wound winding using a single full-width conductor rather than many wire-type conductors per layer. The individual layers are separated by a high-quality kraft paper insulation.

The magnitude of the losses generated by eddy currents is a function of the square of the thickness of the conductor and the square of the frequency of the current flowing in a conductor. Thus, the thinnest wire possible is used. As the wire is made thinner, it must be made wider in order to carry the same current. Carried to the extreme, this results in a foil-thin conductor with a width that equals the height of the winding.

One advantage of a sheet winding is a reduction in voltage stress between the layers of windings. For example, in a design that calls for 10V/turn, and each layer contains 10 turns, there will be:

$$10 \times 10 \times 2 = 200V$$

between the layers.

With a full width sheet winding per turn, the volts between layers will be:

$$10 \times 1 \times 2 = 20V.$$

The selection of the method employed in a transformer is determined by the manufacturer in meeting the specification application requirements.

WINDING MATERIALS

Transformer coils are wound with either copper or aluminum conductors. Some of the principal factors for deciding whether copper or aluminum is used in a transformer are: the size and weight limitations; and the permitted losses. For equivalent electrical and mechanical performance, aluminum-wound transformers weigh less and usually cost less than copper-wound units. Because copper is a better conductor, less material is used for equivalent electrical ratings, and a copper-wound transformer will be slightly smaller than its aluminum counterpart. With both materials, both the winding process and the application of insulation is the same.

Connections to the terminals are welded or brazed; crimp connections are sometimes used. Both copper and aluminum can corrode, which can result in poor bolted contacts. Aluminum oxidizes rather fast unless protective measures are taken, such as using a special grease over exposed parts.

Copper windings are mechanically stronger than aluminum windings and, as a result, will be better able to withstand the forces produced by very high inrush currents produced by frequent starting of motor loads.

Since both types are likely to provide a good service life, most specifiers will accept the manufacturer's standard for the size of the unit. If, however, there is a strong preference for the nonstandard material, it should be noted in the specification. Also, it is possible to specify copper bus terminations for aluminum-wound transformers.

WINDING INSULATION SYSTEMS

Open-wound (dip and bake). Dry-type transformers can have their windings insulated in various ways. One method is to bake the conductor coils and then, when heated, dip them in a varnish with an elevated temperature. This process is an open-wound method and helps ensure penetration of the varnish.

Cooling ducts in the windings provide an efficient and economical way to remove the heat produced by the losses of the transformer by allowing air to flow through the duct openings. This dry-type insulation system operates satisfactorily in many ambient conditions.

Vacuum pressure impregnation (VPI). When greater mechanical winding strength and increased resistance to corona (electrical discharges caused by the field intensity exceeding the dielectric strength of the insulation) is required, vacuum pressure impregnated (VPI) dry-type transformers can be specified. The process forces the insulation (varnish), by vacuum and pressure, into the coils.

Sealed coil. When facing an envi-

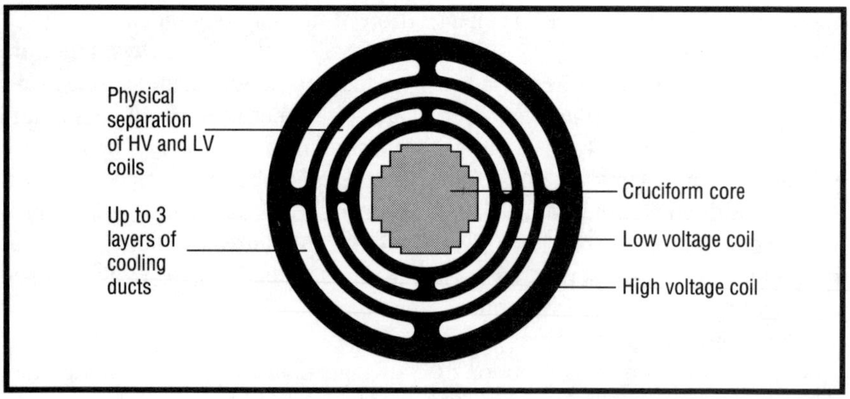

Fig. 10.2. Cross-section through a core and cast-coil windings.

ronment that is too hostile for normal VPI dry-type transformers, the coils are vacuum/pressure filled with a flexible high-temperature varnish. As additional protection, a cast pressure ring is used for clamping the ends of the coil, and the entire core and coil assembly is heavily coated with a special varnish to further protect the unit. This gives these transformers some of the same resistance to mechanical forces, moisture, and contaminants found in cast-coil transformers.

Cast coil. These units have their windings placed into molds and cast, usually under vacuum, in epoxy. The epoxy is a special type that keeps the coils protected from corrosive atmospheres and moisture as well as keeping them secure from the high mechanical forces they may experience as a result of power surges and short circuits. Mineral fillers are sometimes added to the pure epoxy to give it greater strength. Flexibilizers are also added to improve the epoxy's ability to expand and contract with the coil conductors under various transformer load conditions.

Manufacturers use different epoxy filling material and in different amounts. Factors influencing the filler material chosen include:

- temperature rating of the transformer;
- mechanical strength of the coils;
- dielectric strength of the insulation.
- expansion rate of the conductors under various loadings; and
- resistance to thermal shock of the insulation system.

Cast coil transformers consist of separately wound and cast high and low voltage coils, as shown in **Fig. 10.2**. During manufacture, the high voltage winding paper-insulated conductors, are wound over an insulated cylinder. The space between the conductors is filled with fiberglass tape, and a fiberglass cloth sheet is placed over the assembly, onto which the next layer of conductors are wound, and so on until complete.

The completed winding is then placed in a mold sized for the finished diameter of the cast coil. Vacuum filling of the mold with epoxy assures that all the cloth within the windings are thoroughly saturated with epoxy. The mold is then placed in an oven for a number of hours to allow the epoxy to cure and achieve full hardness and strength.

A cast coil insulation system is primarily specified for transformers located in harsh environments. They are better able to withstand heavy power surges, such as frequent but brief overloads experienced by transformers serving transit systems and industrial machinery. These transformers are an alternative to liquid-filled units for harsh environment applications, be-

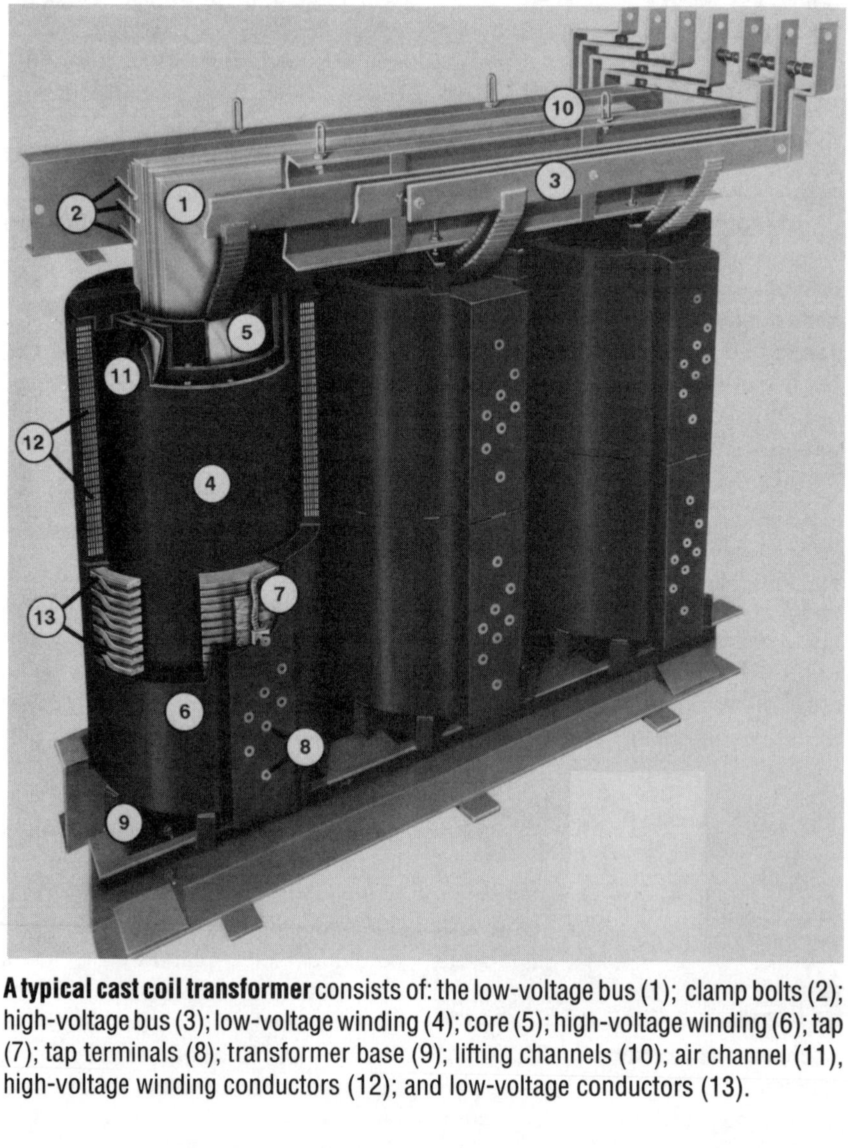

A typical cast coil transformer consists of: the low-voltage bus (1); clamp bolts (2); high-voltage bus (3); low-voltage winding (4); core (5); high-voltage winding (6); tap (7); tap terminals (8); transformer base (9); lifting channels (10); air channel (11), high-voltage winding conductors (12); and low-voltage conductors (13).

cause cast coil units can be constructed to the same high levels of BIL as liquid-filled units while providing ample protection of the coils and the leads. They also operate with lower losses and require less main-tenance than oil-filled types. Also, cast coil transformers, if operated properly, will have a long life expectancy.

Careful attention must be paid to the application, location, and expected atmospheric conditions at the jobsite when writing the specification for a dry-type or cast-coil transformer. The consideration must take into account the likelihood that sufficient maintenance will be available to assure that the air passages through the windings do not become blocked.

STANDARD RATINGS

It is necessary for the specifier of dry-type transformers to include details of their requirements in the specifications. Dry-type power transformers come in such a wide variety of ratings that is difficult, if not impossible, to list them all. Here are some of the most important that must be specified.

Low-voltage encapsulated. Single-phase transformers in the $\frac{1}{2}$ to $37\frac{1}{2}$ kVA range, and 3-phase in the 3 to 45kVA range, are available as wall-mounted totally encapsulated units. They have the advantage of being applicable indoors or outdoors. Primary voltage is limited to 600V maximum and the secondary is usually 120/240V single-phase and many 3-phase voltages, such as 480Y/277, 208Y/120, 240, and more. Impedance and BIL are not items that are required to be specified in this range of product.

Low-voltage floor mounted, ventilated. These are the units most often selected to feed lighting and small motor and appliance loads. They are for installation indoors, but can also serve outdoors if a weathershield is specified. Comparable pad-mounted units are also available. The transformers are generally sized 15 to $37\frac{1}{2}$ kVA single phase, and 15 to 500kVA 3-phase (some up to 1000kVA). These units have virtually the same voltage ratings as the encapsulated ones, and impedance and BIL are generally not included in the specification.

Low-voltage floor mounted, nonventilated. The same 3-phase units can be specified as nonventilated when the units are to be installed in dusty locations, or where the atmosphere may tend to deteriorate the windings. They can be specified in the same ratings, and 80 and 115°C rise units are also readily available.

Higher-voltage, ventilated dry-type. 3-phase transformers in NEMA 1 or weatherproof construction and in approximately the same kVA range as low-voltage ones are available for primary voltages in the 1.2 to 34.5kV class. The BIL depends on the voltage class, the 1.2kV having a BIL of 10kV; 2.4kV having a BIL of 20kV; 5kV a BIL of 30kV; 8.6kV a BIL of 45kV; and 15kV a BIL of 60kV. Similarly, single-phase units from 50 to 500kVA are available in 5kV, 8.6kV, and 15kV classes with similar BILs. Impedance of these units are normally 5.75% for units over 500kVA.

Large higher-voltage, ventilated dry-type. These 3-phase units for

Voltage class (kV)	BIL
1.2	10
5.0	30 or 60
8.7	45 or 75
15.0	60 or 95
25.0	125
34.5	150

Fig. 10.3. Typical values of BIL for medium-voltage ventilated dry-type transformers.

A dry-type transformer used for feeding lighting, receptacles, and small motor loads.

kVA	1.2 kV	5 kV	8.6 kV	15 kV	25 kV
300-500	4.0-5.0	4.5-5.5	5.0-6.0	5.5-6.5	6.0-7.0
501-1000	5.0-6.0	5.5-6.5	5.5-6.5	5.5-6.5	6.0-7.0
1001-1500	—	5.5-6.5	5.5-6.5	5.5-6.5	6.5-7.5
1501-2000	—	5.5-6.5	5.5-6.5	5.5-6.5	6.5-7.5
2001-3000	—	6.0-7.0	6.0-7.0	6.0-7.0	6.5-7.5
3001-3750	—	6.0-7.0	6.0-7.0	6.5-7.5	7.0-8.0
3751-5000	—	6.0-7.0	6.0-7.0	6.5-7.5	7.5-8.5
5001-10,000			No Standard		

Fig. 10.4. Typical impedance values for medium-voltage ventilated dry-type transformers.

Nominal System Voltage	HV BIL	LV BIL	Impulse Tests		
			Chopped Wave		Full Wave
kV	kV	kV	kV Crest	Min. Time to Flashover (Microseconds)	kV Crest
1.2	—	10	10	1.0	10
2.4	45	20	45	1.0	45
4.16	60	30	60	1.0	60
8.32	75	—	75	1.25	75
13.8	95	—	95	1.5	95

Fig. 10.5. Typical BIL values of dielectric tests for winding insulation for cast-coil medium-voltage transformers.

Cooling fans (bottom of enclosure) can be used to boost the rating of dry-type transformers

stand-alone or secondary substation use cover a wide range. They come is sizes to 10,000kVA and operating voltages up to 34.5kV. BILs and impedances vary and can be specified by the purchaser. Typical values are shown in **Figs. 10.3** and **10.4**.

Large higher-voltage, cast or sealed coil. Available in sizes from 500 to 10,000kVA and voltages up to 15kV, have high BIL ratings, ranging from 45kV for 2.4kV class to 95kV for 34.5kV class. Standard impedances are 5.75% for transformers up to 2500kVA, and 5.5% for those from 3750 to 5000kVA, and as specified for larger units. Typical dielectric test values are shown in **Fig. 10.5**.

COOLING

Like liquid-filled transformers, the heat generated in the windings must be dissipated. The spacers inserted between the layers of dry-type transformer windings provide channels through which air can circulate.

Similarly, the kVA rating of dry-type transformers can be raised by the addition of forced air (FA) cooling fans. The general rule is that the addition of fans will boost the rating by $33^1/_3$%. In cast-coil units the increase in output kVA covers a wide range: from $33^1/_3$% to 50%, depending on kVA rating and the manufacturer's standards. The rating increase with fan cooling is usually shown in the specification as two ratings: one without, and one with fans. For instance 2000/2667kVA indicates a $33^1/_3$% increase with fan cooling.

If such a feature is desired, the specification should include the requirement for cooling fans, or as an alternate, that provision be made for their later addition. These provisions include providing all current-carrying parts with sufficient capacity for the fan-cooled rating, and provisions for the addition of an overload relay to control the fans. It should be noted that this provision is sometimes a standard feature that is built into the transformer.

Another way to achieve higher-than-rated capacity without using fans is to use a transformer insulation system that is capable of operating at a higher temperature than could be reached at the hottest point in the unit under full-load conditions at rated kVA. For instance, an 80°C rise transformer would normally allow an additional 30°C maximum gradient in the windings (hot spot) plus have the ability to function properly in a maximum ambient temperature of 40°C, resulting in a maximum operating temperature of 150°C at full load. That is the temperature that the insulation system is designed for.

If a dry-type 80°C rise transformer uses dielectric materials rated for a maximum 220°C operating temperature (the maximum temperature the insulation system is considered to be suitable for), it is possible to "push" the loading of the transformer in excess of rated, up to the point where the temperature of the 80°C rise unit matches that of a 150°C rise transformer. This would result in an overload rating of about 30% more kVA than the 80°C rise rating.

Almost all air-ventilated dry-type 80°C rise transformers use a 220°C class insulation. After making sure that a transformer has a 220°C insulation system, a 80°C rise unit can be operated as if it was a 150°C rise transformer to obtain greater capacity, but at a price. The transformer's losses increase substantially because the conductors are sized for 80°C operation and when additional current is carried, the losses increase as the square of the current (I^2R). And, the life expectancy of the insulation system is reduced, matching that of a 150°C rise transformer. All other factors being equal, a lower temperature rise transformer will have a longer life expectancy when operated within its nameplate rating provided that higher temperature insulation materials were used.

REGULATION

The difference between a transformer's secondary no-load and full-load voltages is a measure of the transformer's regulation. It can be determined as follows:

Regulation (%) =
$[100 (V_{nl} - V_{fl})] \div V_{nl}$

where V_{nl} = no-load voltage
V_{fl} = full-load voltage.

Poor regulation means that as the load increases, the voltage at the secondary terminals drops substantially. This voltage drop is due to resistance in the windings and leakage reactance between the windings.

Voltage regulation and efficiency are improved with low impedance but the potential for serious damage goes up. Specifying a transformer that has good regulation (4 to 6%) is desirable; however, exceptionally good regulation (approximately 2%) may lead to some serious problems. Such a transformer is likely to be severely damaged if a short circuit occurs on the electrical system, especially if the total power on the system is large.

Mechanical stresses in a transformer vary approximately as the square of the current. Stresses resulting from a short circuit could be approximately six times as great in a transformer having 2% impedance as those in one having 5% impedance (where reactance is the major component of the impedance voltage drop). Moreover, lower impedance may require higher interrupting capacity for overcurrent devices downstream, increasing the cost of the secondary switchgear. Of course, a transformer can be built to withstand the high electrical and mechanical stresses and, with a good circuit protection scheme, the problems associated with short circuits can be substantially reduced.

SOUND LEVEL

A transformer, when energized, produces an audible noise. There are no moving parts in a transformer, the sound source is the core. In the presence of a magnetic field the core laminations elongate and contract. These

Equivalent	Self-Cooled Ventilated Voltage Line-to-Line		Self-Cooled Sealed	
Two-Winding kVA	Up to 1.2kV	Above 1.2 kV Up to 15 kV	Above 15 kV Up to 25 kV	Up to 15 kV
0-9	40	45	49	45
10-50	45	50	54	50
51-150	50	55	58	55
151-300	55	58	61	57
301-500	60	60	63	59
501-700	62	62	65	61
701-1000	64	64	67	63
1001-1500	65	65	68	64
1501-2000	66	66	69	65
2001-3000	68	68	71	66
3001-3750	71	71	74	68
3751-5000	73	73	76	70

Fig. 10.6. Typical sound levels of medium-voltage 3-phase ventilated dry-type transformers.

periodic mechanical movements create sound vibrations with a fundamental frequency of 120 Hz and its harmonics. Manufacturers have extensive data on sounds produced by their transformers and they can usually determine the sound level for a particular design quite accurately.

Sound level should be one of the factors to be considered when specifying a transformer. The sound it produces must be under a certain decibel level. Dry-type transformers have sound levels that vary according to the type. **Fig. 10.6** shows some typical values. If the location in which the transformer is known to be sound sensitive, then the exact figures applicable to the transformer must be researched, and a reduced level specified if needed.

Often the location in which a transformer is placed has much to do with the *perceived* sound as does the actual decibels generated. Locating a unit at the end of a long, narrow room, or in the corner of a room can cause a megaphone effect and amplify the transformer's sound. Mounting the unit on a platform that has less mass than the transformer will make the platform serve as a sounding board. Even mounting the unit a distance that is an exact multiple of the 120 Hz wavelength from a solid reflective surface may reinforce the sound waves causing the transformer to seem louder than it actually is.

OPTIONS AND ACCESSORIES

There are may alternative features, besides fan cooling and others already mentioned, that can be specified. Their number is great because of the range of products that go by the name of "dry-type transformers." Accessories are usually an added cost and are sometimes installed while the transformer is being built. Therefore it's advisable to incorporate them in the transformer specification. Following are some of the major ones, but it is far from being a complete list.

Shielding. Depending upon the loads being served, the ability of a transformer to attenuate noise and transients can be helpful. If transients that cause malfunctioning of sensitive electronic equipment cannot be blocked at its source, corrective measures can be taken elsewhere. One method is the use of a shielded isolation transformer, where an electrostatic shield is installed between its primary and secondary (and sometimes also around the individual primary and secondary coils). This type of construction is usually considered when a dry-type distribution transformer serves sensitive electronic equipment.

If common mode disturbances may be a problem on a system where a transformer is to be installed, information on the unit's attenuation ratio should be obtained prior to writing the specification.

Taps. Tapped turns in the winding are used to vary the turns ratio and adjust the voltage to compensate for loads or incoming voltage variations.

Most transformers have four or six 2 1/2% taps (two above and two or four below normal voltage). Normally, taps on dry-type transformers are changed only when the winding is deenergized.

Taps are in many cases standard features; it may only be necessary to specify requirements that are different from standard. This should be confirmed before the spec is written.

Seismic qualified. Units with special bracing are available for units that are to be installed at locations subject to earthquakes. The specification should contain details of the seismic zone involved.

Baffling. When pad-mounted transformers are exposed to the public or to unusual environmental conditions, baffling can be added to keep out driven rain or snow, rodents, or inserted objects. An alternative is outdoor ventilated lockable enclosures to contain the equipment.

Switching. Loadbreak switches or angle connectors can be installed in the transformer cabinet or a closely coupled cabinet. Key interlocks or padlocks to coordinate opening of enclosure panels for operation of the switches are also available.

Finishes. Special paint/finishes for corrosive atmospheres and ultra-violet light degradation and protective provisions for humid environments can be specified. In rare cases, a stainless steel cabinet can be specified for extra corrosion protection;

ACCEPTANCE TESTING

Before accepting a transformer its important that it meet a series of tests to make sure that the unit will function as specified. It's not usually required that tests be performed on low-voltage transformers, but when the units operate at the medium-voltage levels it is recommended that a number of tests be carried out and that these test comply with applicable ANSI, IEEE and NEMA standards. When special testing is being carried out, a transformer manufacturer will allow a purchaser or a representative, such as a consulting engineer, to witness the tests. The usual standard transformer tests that are performed for each unit include:

- resistance measurements of all windings (for calculating winding temperature and I^2R component of winding losses);
- ratio (for voltage relationship);
- polarity & phase relationship for 3-phase units (important when two or more transformers are to be operated in parallel);
- excitation current (to verify core design is correct and relates to efficiency);
- no-load core loss (to verify core design is correct and relates to efficiency);
- impedance (via short circuit test. Magnitude of current determines needed breaker capacities and/or fuse size and is needed for coordinating relaying schemes);
- load loss (as with no-load losses, directly relate to the transformer's efficiency);
- regulation (to determine voltage drop when load is applied);
- temperature (to help assure design limits will not be exceeded); and
- applied potential and induced potential (to verify dielectric strength)

Additional tests that may be applicable, depending upon how and where the transformer will be used, could include:

- impulse (where lightning and switching surges are prevalent);
- sound level measurements;
- insulation power factor (for future reference to help determine aging process of the insulation); and
- short-circuit calculations.

Each test has a particular objective that helps determine a transformer's suitability for use. It is important to work with the transformer manufacture in establishing what tests should be carried out. Additional tests are an added cost to the unit. When specifying additional tests, the experience of the manufacturer, including test data from essentially duplicate units, should not be overlooked. When this data can be utilized, the extra cost to carry out tests may be able to be avoided without sacrificing the quality of the transformer.

RECEIVING AND INSTALLING TRANSFORMERS

11

Power transformers are large, bulky, heavy items with few, if any, moving parts. They appear so rugged that little attention is paid to the paperwork and brochures that come with them. Aside from a pad or mounting bracket, they seem to place few demands on the skill of the installer.

From the maintenance aspect, they just sit there and hum. So they also tend to be ignored once they are in operation.

Both of these approaches are wrong! Since they are so critical to the distribution of power within a facility, transformers merit full attention during installation and operation. By following some simple procedures, the transformer can give many trouble-free years of service.

RECEIVING

Transformers should be carefully examined for damage when they arrive at the jobsite. Special attention should be given to the bushings and tank-mounted accessories of liquid-filled units.

Storing. If the transformers are to be stored before installation and energizing, they must be protected against condensation and weather. Dry-type transformers must be stored in dry, clean, heated areas. If there is any possibility that the windings may have absorbed moisture, the transformers must be dried using blowers, heating elements, or incandescent lamps in the enclosure, before the unit is energized.

Assembly. Liquid-filled transformers of moderate size are shipped assembled and filled with oil or other insulating liquid. Large units may be shipped disassembled, with bushings and radiators removed, and filled with inert gas. These must be assembled and filled with the insulating liquid. Assembly and filling must be done with great care in accordance with the manufacturer's instructions. Filling must be done under vacuum to prevent entry of air and moisture. More about this subject can be found in Chapter 12 of this book.

Moving. Transformers are heavy and usually require overhead cranes for lifting into final position. Lifting hooks are provided, and spreaders should be used to prevent damage. Somewhat smaller units can be moved into place using pipe of conduit under the unit. An air-cushion device is also available for medium weight units. Dry-type transformers are especially top-heavy. Smaller units can usually be handled with fork lifts.

Transformers must be moved in the upright position to avoid tilting or tipping.

INSTALLATION

Outdoor units should be installed on a level concrete pad. Indoor units should be installed on a level concrete floor and located carefully with respect to reflecting surfaces to minimize noise.

Location. When locating a dry-type transformer, it is essential that adequate ventilation be provided to ensure proper cooling. Thus, it must be located so that ventilating openings are not blocked by walls or other obstructions. This is not only good practice, it is also a requirement of Article 450 of the NEC. The NEC states that the required clearances be clearly marked on the transformer.

When transformers are located in vaults, electrical closets, and other restricted places, sufficient ventilation should be available to maintain the air temperature within established ambient limits when measured near the transformer air inlets.

Connecting. Liquid-filled transformers have bushings for primary and secondary connections located in air-filled terminal chambers. Some larger transformers have exposed primary bushings. Medium-voltage cable connections require voltage stress relief similar to that required for medium-voltage switchgear. Secondary connections may be cable or busway. All bolted terminations and connections should be tightened to their recommended torques.

Dry-type transformers must be connected carefully to prevent damage to the conductors from hot air rising through the coil structures. Small units have connections near the bottom. The connections should be made to the proper tap required to assure that the operating voltage at the load is within specified limits. Cables should be kept in the lower part of the enclosure, as marked, and must be used at the 60°C ampacity (under 100A) or 75°C ampacity (100A and above), regardless of the insulation rating. Larger units may have connections located higher than the coils; 90°C conductors should be used, sized for their 75°C ampacity.

Secondary connections of higher ampacity may be for busway rather than cable. All connections and terminations must be tightened to the recommended torques.

ENERGIZING

The characteristics of the different types of transformers really come into play when preparing to place the transformer into operation.

Dry-type transformers should be cleared of dirt and debris and inspected to ensure that the flow of cooling air is not obstructed. The core and coils are mounted to the frame on vibration isolators. To prevent damage to these isolators during shipment, shipping blocks or other supports are bolted in place. They are usually tagged and marked "REMOVE BEFORE ENERGIZING." The manufacturer will indicate the procedure to remove the shipping supports. This must be done to permit the vibration isolators to function; otherwise, the transformer will be noisy. The covers of the enclosure should be installed and tightened so that they will not vibrate.

Liquid-filled transformers. Before energizing it is essential to confirm that the insulating liquid is within specified tolerances. Normally, the tests outlined in Chapter 12 are performed. The values obtained should be recorded; they serve as a baseline for future maintenance testing of the liquid..

The external tap changer of a liquid-filled unit should be set and padlocked to the tap that will assure proper voltage at the load. If this is unknown before energizing for the first time, it is possible to do so at a later time if a load tap switch has been specified. If a deenergized load tap changer is employed, then a shutdown will be necessary.

All accessories should be checked for proper operation, and alarm initiating devices set and tested. This last item is of extreme importance. It is too late to discover a faulty alarm circuit after the resulting damage has been done. Yet, this simple step is often ignored.

NEC INSTALLATION REQUIREMENTS

All electrical equipment has sets of rules that govern how it is to be installed and protected; transformers are no exception. The following are extracted from Article 450 of the NEC. It would be a good idea to include these items in specifications for the design and installation of systems in which transformers are located. For references to specific sections of the NEC, see EC&M's book entitled *Understanding NE Code Rules on Transformers*.

Transformers and transformer vaults must be readily accessible to qualified personnel for inspection and maintenance. This general rule, however, has two exceptions that allow

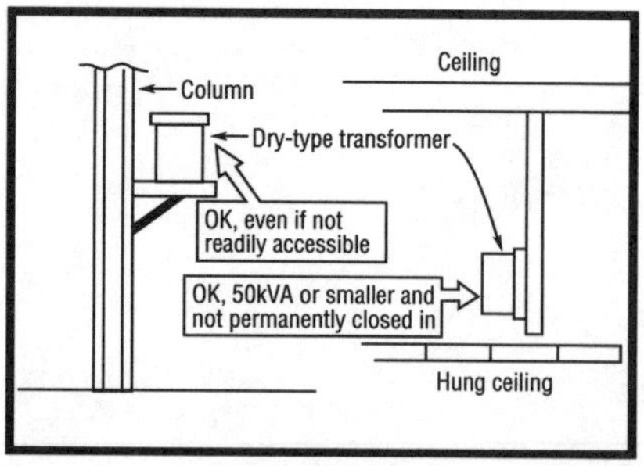

Fig. 11.1. Although these transformers are not "readily accessible," these locations are permitted by the NEC.

NEC rules allow dry-type transformers operating at 600V or less to be mounted on columns.

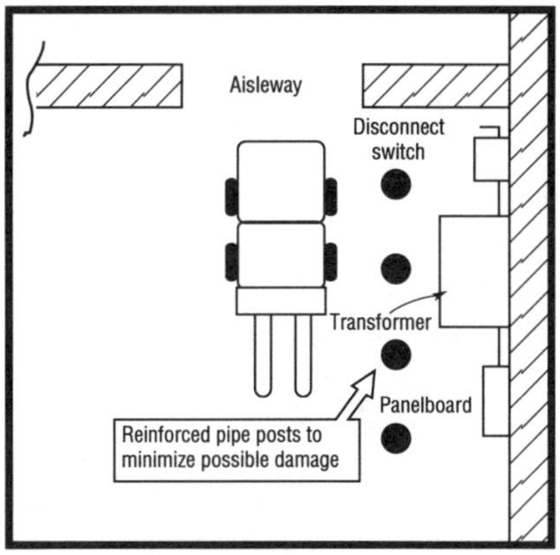

Fig. 11.2. Transformers must be protected from physical damage.

dry-type transformers operating at a voltage not exceeding 600V to be located in positions where they are not readily accessible when they are:

- located in the open on walls, columns, or structures; or
- installed in fire-resistant (minimum fire rating of 1 hr) hollow spaces of buildings not permanently closed in by structure, provided they meet the ventilation requirements of **Sec. 450-9**, and are rated at 50kVA or less.

Location. Transformers and transformer vaults must be readily accessible to qualified personnel for inspection and maintenance. According to the NEC definition of the term "readily accessible," the transformer can be located on a balcony or other location as long as a permanent ladder of stairwell provides ready access to the equipment. Having to use a portable ladder does not qualify as providing ready access.

There are, however, some exceptions to this rule. As shown in **Fig. 11.1**, the transformer is permitted to be located in the open on walls, columns, or structures.

Location of a dry-type transformer on a column or on a support structure tied to a wall or ceiling is permissible because there is very little need to service such equipment. A liquid-filled transformer, however, is not allowed to be mounted where it is not readily accessible because the need for servicing is considerably greater.

In addition, it is allowed to be installed in a fire-resistant (minimum fire rating of 1 hr) hollow spaces of buildings, not permanently closed in, provided the ventilation within the space meets the needs of the transformer. If the dry-type transformer is above a solid hung ceiling, it would not be in the open and would not qualify for the exception.

An additional requirement is that the area around a transformer must be kept clear of stored material that would impede access to the unit.

Mechanical protection. Appropriate provisions must be made to minimize the possibility of damage to transformers from external causes where the transformers are exposed to physical damage. Aisleways within a facility that provide access for forklift and other vehicles present particular dangers to transformers and other equipment mounted nearby. Provisions *must* be made to limit the possibility of damage. As shown in **Fig. 11.2**, this can be accomplished by posts located in front of the transformer, bumper strips, fences, or other means.

Exposed energized parts. All exposed energized parts (such as the bushings of liquid-filled transformers, and switches) must be guarded. In addition, entrances to rooms or enclosures in which they are located must have prominently displayed warning signs forbidding unqualified persons from entering. In addition, the operating voltage of exposed live parts of

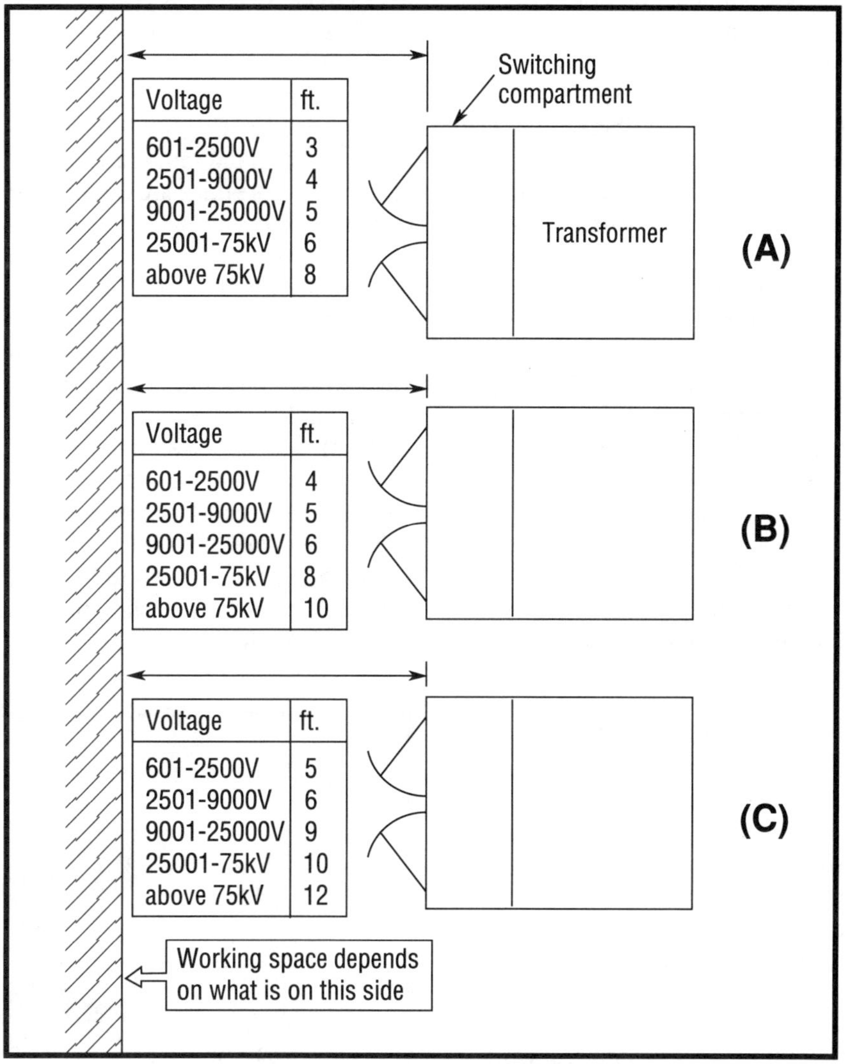

Fig. 11.3. When installing a transformer operating at a medium or high voltage, the NEC requires that adequate work space is allowed in front of energized live parts. There clearances depend upon the conditions that apply. In **(A)**, there are no live parts on the other side; in **(B)** there are grounded parts on the other side, or it is a concrete, brick, or tile wall; and in **(C)** there are exposed live parts on the other side of the work aisle.

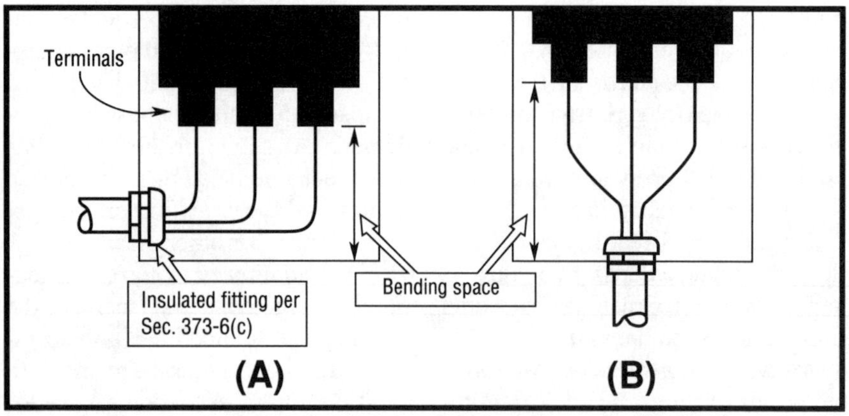

Fig. 11.4. It is important to check the cable bending space within a transformer terminal compartment before deciding where the cables should enter.

transformer installations must be indicated by signs or visible markings on the equipment or structures.

600V, nominal, or less. The approved methods of guarding are:

- location in a room, vault, or similar enclosure accessible only to qualified persons;
- partitions or screens so arranged that only qualified persons will have access to the space within reach of the live parts. Any openings in these barriers must be sized and located so that persons are not likely to come into accidental contact with the live parts or to bring conducting objects into contact with them;
- location on a suitable balcony, gallery, or platform so elevated and arranged as to exclude unqualified persons; or
- by installing at an elevation of 8 ft (2.44 m) or more above the floor or other working surface.

Over 600V, nominal. Placement of this equipment is already governed by other NEC rules, so the main emphasis here is on assuring that there is adequate working space allotted from live parts (such as switches within a compartment of padmounted transformers) and what is on the other side of the work aisle. As shown in **Fig. 11.3** (previous page) there are three possibilities, the required working space varying accordingly. In **(A)**, the other side contains no exposed live parts or they are guarded by insulating materials. In **(B)**, the other side contains grounded parts or concrete, brick, or tile walls (which are considered to be grounded surfaces. In **(C)**, both sides contain exposed live parts (such as another switching compartment opening onto the same space).

Other requirements include that where switches, cutouts, or other equipment operating at 600V or less are installed in the same room or enclosure as where there are live parts of transformers operating at higher voltages, the two must be separated by a partition, fence, or screen. An exception to this requirement is made for LV equipment that only serves equipment within the room or enclosure.

Unguarded live parts (such as bare busbars) that is above the working space must be maintained at a minimum elevation or $8^1/_2$ ft if the nominal voltage between phases is from 1001V to 7500V; 9 ft if from 7501 to 35000V; and (9 ft plus 0.37 inches per kV above 35kV) for any voltage above 35kV.

Making connections. A minimum wire bending space is required at the line and load terminals of 600V and below transformers. This requirement is of importance to the installer of the equipment because the intent is to assure that adequate space is provided at the terminals to allow pulling in the conductors without overcrowding the space, and also that adequate gutter space is available for offsetting the conductors so they can be connected to the terminals without exceeding the allowable bending radius of the conductors and their insulation. The installer must not deflect the conductors within the transformer enclosure unless it has an adequately sized gutter.

There are two conditions that have to be taken into account when planning how to enter the termination compartment of the transformer. Depending upon the location of the entrance, two different tables of required clearances apply as shown in **Fig. 11.4**. It is, therefore, important to check the distance from the terminals to the wall of the compartment before deciding where to enter with the cables.

The volume requirement for a wiring compartment of a transformer in which pigtail connections are to be made is completely different from that for a transformer with terminals. A separate table listing volume requirements is provided for this case.

The NEC also specifies the use of an insulated fitting where ungrounded No. 4 or larger conductors enter the wiring gutter of a transformer. The

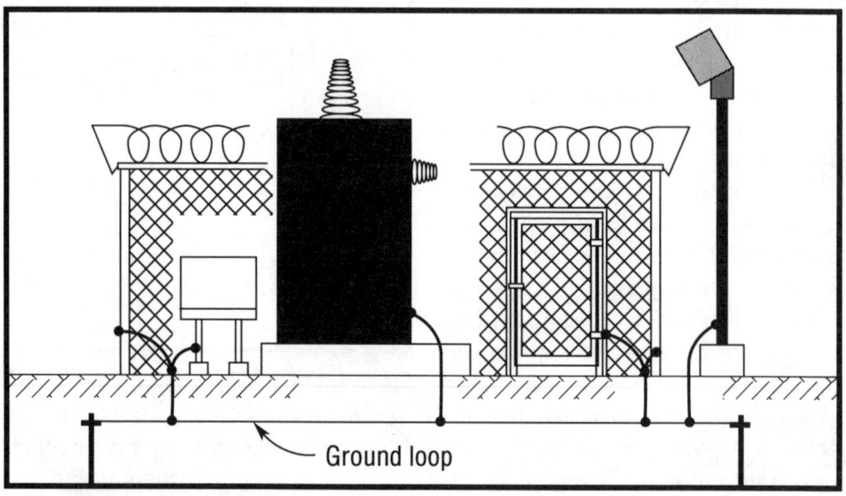

Fig. 11.5. All noncurrent-carrying metallic parts associated with a transformer installation must be grounded.

fitting must provide a smoothly rounded insulating surface, unless the conductors are separated from the raceway fitting by substantial insulating material that is securely fastened in place. An exception to this rule is made where threaded hubs or bosses that are an integral part of an enclosure provide a smoothly rounded or flared entry for conductors.

Conduit bushings made wholly of insulating material must not be used to secure a raceway. The insulating fitting or insulating material must have a temperature rating not less than the insulation temperature rating of the installed conductors.

Grounding. Exposed noncurrent-carrying metal parts of transformer installations, including fences, guards, and similar items must be grounded where required under the conditions and in the manner specified for electric equipment and other metal parts in Article 250.

Not only must the transformer casing and be grounded, but also any equipment, fencing, or other metallic items associated with the transformer installation, as illustrated in **Fig. 11.5**.

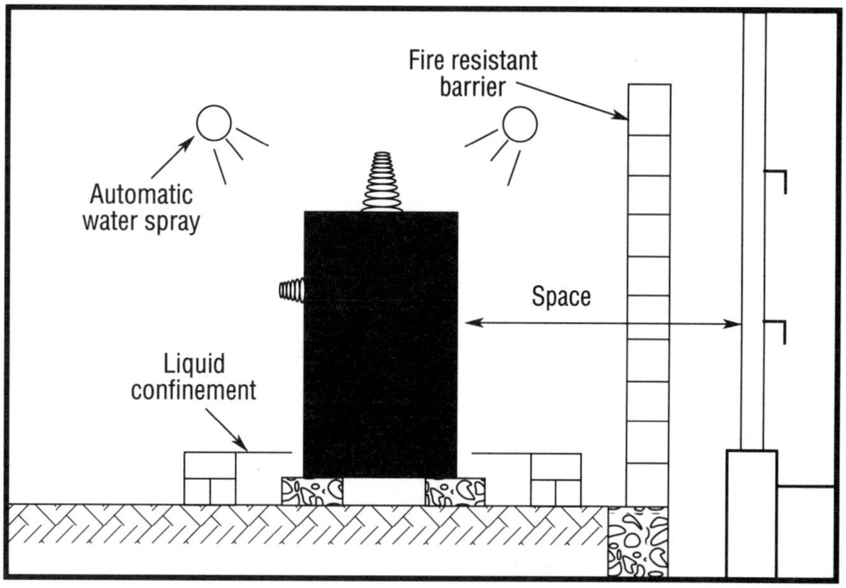

Fig. 11.6. Safeguards that can be used to protect a structure from damage due to a transformer oil fire.

NEC PROVISIONS FOR SPECIFIC TYPES OF TRANSFORMERS

Transformers vary greatly in size and ratings. For this reason, not all the rules dealing with this equipment can apply to all units. The NEC, thus, spells out specific provisions that are applicable to the different types of transformers.

Dry-type transformers not over 112 ½ kVA and installed indoors must have a separation of at least 12 inches (305mm) from combustible material. There are two exceptions to this rule.

- Where separated from the combustible material by a fire-resistant, heat insulating barrier.
- If the transformer is rated at 600V nominal or less, and is completely enclosed without ventilating openings.

Larger kVA dry-type transformers must be installed in a transformer room of fire-resistant construction. There are also two exceptions to this location rule that apply to transformers with 80°C rise or higher ratings.

- If separated from combustible material by a fire-resistant, heat-insulating barrier, or separated by not less the 6 ft (1.83 m) horizontally and 12 ft (3.66 m) vertically.
- If completely enclosed except for ventilating openings.

Dry-type transformers operating at a voltage higher than 35kV must be installed in a vault complying with the requirements outlined in Chapter 8 of this book.

Dry-type transformers installed outdoors must have a weatherproof enclosure. Transformers rated at over 112½ kVA must not be located within 12 inches (305 mm) of combustible materials or buildings. An exception is made for transformers having 80°C rise or higher ratings, and for completely enclosed types (including those with ventilating openings).

Liquid-insulated transformers with listed less-flammable liquids are permitted to be installed without a vault in Type I and Type II buildings in areas in which no combustible materials are stored, provided:

- there is a liquid confinement area;
- the liquid has a fire point of not less than 300°C; and
- the installation complies with all restrictions provided for in the listing of the liquid.

Indoor transformer installations not meeting the restrictions of the liquid listing, or installed in other than Type I or Type II buildings, or in areas where combustible materials are stored must be provided with an automatic fire extinguishing system and a liquid confinement area; or be installed in a vault. Transformers installed indoors and rated over 35kV always must be installed in a vault.

Less-flammable liquid-insulated transformers are permitted to be installed outdoors attached to, adjacent to, or on the roof of Type I or Type II buildings. Where the transformer is adjacent to combustible material, fire escapes, doors, or windows, it must also have safeguards such as a fire barrier; and/or space separation; and be in compliance with all of the requirements provided for in the listing of the liquid.

Oil-insulated transformers, when installed indoors, must be installed in a vault. There are exceptions to this rule, including the following.

- If less than 112½ kVA, the transformer vault can be constructed of reinforced concrete not less than 4 inches (102 mm) thick.
- If not over 600V and the total capacity at one location does not exceed 10kVA in a section of the building classified as combustible, or 75kVA where the surrounding struc-

ture is classified as being fire-resistant, a vault is not required provided that suitable arrangements are made to prevent a transformer oil fire from igniting other materials.

• If in a detached building, the construction does not have to meet the specifications for a vault if neither the building nor its contents present a fire hazard to any other building or property; the building is used only to supply electric service; and the interior of the building is accessible only to qualified persons.

When oil-insulated transformers are located outdoors, combustible material, combustible buildings and parts of buildings, fire escapes, and door and window openings, must be safeguarded from fires originating in transformers installed on roofs, attached to, or adjacent to a building or combustible material.

As seen in **Fig. 11.6**, safeguards are defined as being space separations, fire-resistant barriers, automatic water spray systems, or enclosures that confine the oil of a ruptured transformer tank. One or more of these safeguards must be applied according to the degree of hazard involved in cases where the transformer location presents a fire hazard.

Oil enclosures can be fire-resistant dikes, curbed areas or basins, or trenches filled with coarse crushed stone. Oil enclosures must be provided with trapped drains where the exposure and the quantity of oil involved are such that removal of oil is important.

Modification of transformers in an existing installation that change the type of the transformer, such as retrofilling with a different type of insulating medium, the transformer must be marked to show the type of insulating liquid installed. The modified transformer installation must comply with the applicable requirements for that type of transformer. For instance, if an askarel-filled transformer were retrofilled with oil, an entirely different set of rules will apply because of the different degree of fire hazard the transformer would pose.

TRANSFORMER MAINTENANCE

12

A proper maintenance program can prevent a majority of transformer failures. The value of a maintenance program varies depending on the impact of a service failure. Maintenance planners should perform a study to evaluate this impact and then define a maintenance program that meets their particular needs. This impact study should concentrate on the costs of equipment replacement and service interruption, safety, and the importance of reliability to the particular installation.

ECONOMIC CONSIDERATIONS

The economic value of regular maintenance can be equated to the replacement and consequential costs that would be avoided minus the cost of conducting a maintenance program. The cost of replacing or repairing a transformer on an emergency basis is invariably higher than the cost of a planned installation or repair. The consequential costs due to the loss of power and the subsequent business interruption can vary substantially and should be evaluated carefully.

Safety concerns are even more important than any financial considerations. These include the hazards to the public and operating personnel due to the direct effects of a failure such as fire, explosion, or arcing, and indirect effects such as the loss of alarm circuits or critical processes that might result from a sudden power interruption. Indirect effects are usually less obvious than the direct effects but are often important and deserve careful consideration.

A transformer providing power to a hospital, where untimely service interruptions are intolerable, would be at the more critical end of the scale and should have an intense maintenance program on the basis of reliability alone. A transformer protected from public exposure, feeding a residential or office lighting load and accessible for ready replacement would be at the less critical end of the scale. In these circumstances it might be allowable to minimize maintenance requirements if the economic considerations are acceptable and safety will not be compromised.

A typical schedule for transformer maintenance testing is shown in **Fig. 12.1** (next page).

TESTING PROGRAMS

There are standards and guides that give recommendations for maintenance programs and tests. ANSI/NFPA 70B, *Recommended Practice for Electrical Equipment Maintenance*, includes suggested maintenance programs and describes the individual tests. ANSI/IEEE C57.125, *Guide for Failure Investigation, Documentation and Analysis for Power Transformers and Reactors*, includes discussion of evaluation tests and gives guidelines for their use and interpretation. Manufacturers' instruction books may contain some references to particular maintenance requirements but they do not attempt to outline complete maintenance programs.

As was discussed in Chapters 9 and 10, acceptance tests are routinely carried out by the manufacturer of large power transformers before shipping. They can be witnessed by the purchaser, but what is important is that the data taken at that time be obtained and placed in a maintenance file. These values will serve as the basis for evaluating results of later testing once the transformer has been put into service.

TESTS BEFORE ENERGIZING

Before initial energization, if there is no evidence of shipping damage, the tests should include an internal inspection, winding insulation resis-

Category	Elements	Frequency		
		Low concern	Average concern	High concern
Routine tests	•Record temperature •Record pressure •Record liquid level •Record load/time •Record ambient temp. •Record visual leaks •Record tank heating •Record sound level (norm/abnormal)	Monthly	Weekly	Daily
Periodic tests	•Oil samples •insulation p.f.[1] •Insulation resistance •Core ground[2] •Winding ratio •Gas-in-oil analysis[3]	5-yr. intervals	1-yr. intervals	3- or 6-month intervals
Acceptance tests	•Oil samples •Internal inspection •Insulation resistance •Core ground •Winding resistance •Excitation current •Impedance •Insulation p.f.			

Fig. 12.1. Typical schedule for transformer maintenance.

tance measurements (megohm measurements), core insulation resistance measurements (if the core is insulated and grounded at a single point), and winding ratio measurements. If the transformer is of the liquid-filled type, in addition a sample should be drawn for dielectric breakdown test and water content analysis. Excitation current measurements, impedance measurements, and winding resistance tests should be included if there is any evidence of shipping damage.

When conducting the initial tests, take time to record all essential details of the tests as they will help immensely when making comparison of results when these tests are duplicated in the future. These details should include the date and time of the tests, identification of the instruments used, description of connections, dielectric temperature if liquid filled, and the names of individuals involved.

ROUTINE MAINTENANCE CHECKS

Visual observations and gauge readings should be performed weekly. However, these inspections may be performed at intervals varying from a month or more to as often as daily depending on the nature of the installation.

Such inspections consist of visual observations and recordings that can be safely made with the transformer in service. These checks can be made on most transformers except for some in confined spaces or inaccessible locations, such as underground residential or pole top transformers. These routine checks are simple to perform but should only be entrusted to skilled observers trained in the hazards of the electrical system and instructed on what data should be recorded. Trends in the readings over time are very important. The readings of different parameters for any single time period, viewed by themselves, are seldom significant.

The load on the transformer, winding temperature (if available), and the ambient temperature are important items to be recorded. Provision should be made for measuring the air temperature near the transformer as this is the ambient temperature of interest.

If load values are available, they should be recorded. However, in most commercial installations, load measurements are not made. For such installations the time of day that the observations were made should be recorded, along with an evaluation of the load as normal, high or low. When low or high loads are indicated, comments should be added identifying the cause of the change (part of process shut down, equipment additions, etc.). These observations will allow comparisons between readings at similar times, ambient temperatures, and load conditions and thus add much value to the data.

If a transformer is equipped with a winding temperature indicator, this instrument will provide an indication of the temperature of the coils. It is impossible to read the winding temperature directly on most transformers.

Visual checks should include inspection for evidence of localized heat damage to the paint, dust accumulation on the cooling surfaces, excessive vibration and sound level. The sound level assessment is very subjective but can have value if the same observer is making the evaluation and is sensitive to sudden pronounced changes in sound level.

PERIODIC TESTS

Periodic testing should be performed on a scheduled basis (usually annually) and requires the electrical isolation of the transformer. Since these tests require special equipment and techniques, it is advisable to rely on an experienced service organization for their performance to ensure consistency of results as well as for safety and efficiency. Also, review the manufacturer's instruction manual for any recommendations on maintenance requirements for the external components.

Winding insulation resistance as measured with a megohm meter is an indication of the strength of this insulation to an applied voltage. The transformer must be isolated electrically from the system with the primary and secondary circuits opened as near to the transformer terminals as possible.

A sensitive bridge or microhmmeter with an output of 500 or 1000V must be used that is capable of measuring in the micro-ohm range in the case of the secondary winding and up to 20 ohms for the primary winding. These values may be compared with original test data corrected for temperature variations between the factory values and the field measurement or they may be compared with prior maintenance measurements.

To conduct the test, all terminals of

each winding should be connected together with light copper wire. There should be no connection between secondary and primary windings. The megohmmeter should be connected with one lead to the terminals of each winding and the insulation resistance measured and recorded. The megohm meter leads should then be connected to take readings between each winding and ground with the untested windings grounded.

As a rule, a transformer is acceptable for service if the megohm value is greater than 1000 megohms for the primary winding and 100 megohms for the secondary winding and all other tests are within acceptable limits. These minimum values would be questionable if prior readings taken under similar conditions are more than 5% higher.

Megohm readings on a liquid-filled transformer will vary from time to time because the instruments apply a DC voltage. Some extraneous components typically found in the oil of an operating transformer will align in a DC field over time and give erroneously low readings. Refer to ANSI/ IEEE C57.125 to determine minimum acceptable megohm values for a particular transformer.

Core ground insulation resistance. Extraneous grounds might result in low resistance circuits around a core leg or yoke that would allow excessive circulating currents to flow. These currents can be high enough to melt core steel and lead to a winding insulation failure. To make this measurement, locate the core ground. This is usually a copper strap from the core laminations secured under a bolt on the steel supporting clamp. The bolt can be loosened at the top clamp and the copper strap slid free. One lead of a megohm meter is then connected to the free strap and the other lead to the bolt on the support clamp and the insulation resistance measured and recorded.

Winding ratio test. Determining the ratio of the transformer windings relates to the number of primary winding turns divided by the number of secondary winding turns. This ratio is useful in determining if there are any shorted turns or open winding circuits. The measured ratio should be within one half of one percent (0.5%) of the ratio of the rated voltages between the windings as given on the transformer nameplate.

Ratio measurements should be recorded for all tap positions on all phases. Sensitive instruments for measuring the winding ratio are available that operate at very low voltage levels and are thus simple and safe to use. It is possible to measure the winding ratio by applying 120V AC or less to the primary winding and measuring the voltage developed on the secondary winding.

In using this method, great care must be taken to measure all voltages as precisely as possible. It is not always possible to obtain the desired accuracy with voltage measurements so the use of specialized instruments that are available for this test is highly desirable.

Insulation power factor. To confirm that the winding insulation is not affected by moisture or other contaminants that would reduce its insulating properties, a measurement of the insulation power factor is sometimes performed. An insulator under the application of an AC voltage will allow a very small leakage current to pass. Part of this current flow is due to the resistive component of the insulation material and part is capacitive due to the dielectric constant of the material. Insulation power factor is the ratio of the resistive current component to the total leakage current under an applied voltage stress.

To carry out an insulation power factor test, the transformer must be electrically isolated and the winding terminals connected together as described under the winding insulation resistance tests. An alternating voltage is applied, in succession, to each winding, and measurements are made by instrumentation that allows the separation of the resistive component of the current. The resistive component is then compared with the total current to determine the ratio of these values or power factor.

A perfect insulator would pass no resistive current when subject to an AC voltage stress. As an insulator becomes less perfect, the resistance current increases. Therefore, the better the quality of the insulation, the smaller the resistive component of the current, and, the lower the power factor. On transformers rated below 35kV, this value should be less than 1%. The external porcelain of the transformer bushings must be free from moisture or contamination as these will give erroneously high readings.

Hot-collar test is often used for testing transformer bushings having a solid conductor, without disconnecting the transformer winding. It is unique because detection of moisture and other forms of deterioration in a section of a bushing or pothead is possible without disturbing the main connections. This is done by applying a collar-shaped electrode over the section of the bushing to be investigated and applying an AC test potential between the electrode and the bushing conductor; the dielectric losses in the bushing insulation under the collar can then be measured.

UST measurement is used on bushings equipped with capacitance or power-factor taps to test the main insulation of a bushing exclusive of connected windings, interrupters, etc. It is used also for direct measurements on interwinding insulation of transformers and rotating machinery.

MAINTENANCE ACCEPTANCE TESTS

Certain acceptance tests may be repeated sometimes in the life of a transformer to demonstrate its serviceability after a period of unusual service or prolonged inactivity. These tests are more extensive than routine tests. The results of the initial acceptance tests that were conducted now provide key information when making a comparison of test results.

Reference to the manufacturer's instruction book, which is usually identified on the transformer nameplate, is advisable to see if there are any particular acceptance test instructions given.

Impedance tests are performed after a transformer has been removed from service for extensive repairs, or has been subjected to an unusual service event, such as a severe fault on the secondary winding. Mechanical distortion of the windings will cause a change in the impedance of the windings that can often be detected by an impedance measurement. To be able to use this test most effectively it is best to make a measurement at initial installation to establish a benchmark value.

An impedance test is performed by electrically connecting the secondary terminals together with a conductor capable of carrying at least 10% of the line current and applying a reduced voltage to the primary windings. The applied voltage is measured at the primary terminals and the current is measured in each line. These measurements must be taken with instruments of known calibration. The ratio of voltage to current should be within 2% for each phase and should not vary more than 2% between tests. A variation of more than 2% indicates the possibility of mechanical distortion of the winding conductors and should be investigated. For additional details on this test refer to ANSI/IEEE C57.125, Appendix A.

Winding resistance. If the possibility of winding damage is suspected, a measurement of the winding conductor resistance should be made. This test requires an ohmmeter capable of accurately measuring resistance in the range of 20 ohms down to small fractions of an ohm. This test must be performed with the measurement made over a time span long enough to allow the readings to stabilize. The initial readings will be high because of the reactance of the transformer winding. When the reading is unchanged for at least 30 seconds, record the indicated value. Winding resistance varies with temperature so the indicated temperature should be recorded. The readings taken should be compared with prior readings, factory test values or readings from identical transformers. On multi-phase transformers, readings between phases should be comparable. Variations of more than 5% may indicate a damaged conductor in a winding.

Excitation current. For the excitation current test the transformer should be disconnected from all external circuit connections and a reduced voltage applied to the primary winding. For most transformers a single phase 120V supply works very well. The voltage is applied to each phase in succession and the applied voltage and current measured and recorded. If there is a defect in the winding or magnetic circuit that is circulating fault current there will be a noticeable increase in the exciting current.

There is normally a difference between the center leg and the two outer legs so it is preferable to have established benchmark readings for comparison. Variations between readings should not exceed 5% when comparing with prior readings and the phases on any single test should compare within 15%.

Applied voltage test. The applied voltage test is more commonly referred to as the "highpot" test. This is performed by connecting all terminals of each individual winding together and applying a voltage between the windings and from each winding to ground in separate tests. Untested windings are grounded during each application of voltage.

Although ANSI/IEEE C57.94 lists the applied voltage test as an optional pre-service or periodic test it should be used with caution as it is a test that can cause insulation failure. It should be regarded as a proof test to be conducted when there has been an event or pattern in the operating history that makes the insulation integrity of the transformer suspect.

ANSI/IEEE C57.94 states that either AC or DC voltage tests are acceptable for applied potential testing but that the DC applied voltage should not exceed the rms value of the standard test level. AC voltage rms values are limited by C57.94 to 75% of the original test levels for initial installation tests and 65% of the original test levels for routine maintenance tests. The original test levels are specified in ANSI/IEEE C57.12.01 and the tests described in ANSI/IEEE C57.12.91. These standards should be carefully reviewed before conducting any applied potential tests. If the original factory test reports are available they should be consulted to determine the original factory test levels.

DC applied voltage tests are often conducted in the field because DC test sets are smaller and more readily available than AC applied voltage sets. With DC tests the leakage current can be measured and is often taken as a quantitative measure. DC leakage current can, however, vary considerably from test to test because of creepage across the complex surfaces between windings and between windings and ground. The use of AC voltage is preferable since the transformer insulation structures were designed, constructed and tested with the application of AC voltage intended.

DRY-TYPE TRANSFORMER TESTS

Maintenance requirement for dry-type transformers, although similar to liquid filled transformers in many ways, differs sufficiently to warrant separate consideration.

ANSI/IEEE C57.12.01, *IEEE Standard General Requirements for Dry-Type Distribution and Power Transformers* groups dry type transformers into the classifications of ventilated, non-ventilated, and sealed. As with liquid filled transformers, a maintenance program on dry type transformers should include routine inspections and periodic checks. Acceptance tests should be performed as the need is indicated by review of the maintenance data and the operating history. The frequency for routine inspections and periodic checks will depend on the transformer classification as well as the operating environment, load conditions, and requirements for safety and reliability. ANSI/IEEE C57.94, *IEEE Recommended Practice for Installation, Application, Operation, and Maintenance of Dry-Type General Purpose Distribution and Power Transformers* covers general recommendations for the maintenance of dry-type transformers and is a valuable reference source.

Ventilated transformers will require more frequent periodic checks than those that are non-ventilated. Sealed transformers will require less frequent periodic checks than either non-ventilated or ventilated.

The frequency of periodic checks will depend on the degree of atmospheric contamination and the type of load applied to the transformer. This is especially true for the non-sealed transformers since the ambient air and any contaminant dust or vapors it carries can contaminate the internal electrically stressed components. As routine inspections are made, the rate of accumulation of dust and moisture on the visible surfaces should serve as a guide for scheduling periodic maintenance.

Routine checks. Dry type transformers do not have the gauges that are needed on liquid filled transformers to monitor temperature, pressure and liquid level, so the routine checks are more subjective and consist mainly of visual and audible observations. Sealed dry type transformers do have pressure gauges and these should be routinely checked. It is imperative to have a complete checklist developed for each transformer with essential observations and data recorded and preserved.

Visually inspect louvers, screens and any visible portions of internal coil cooling ducts for accumulated dust. Do not remove any panel or cover unless the transformer is deenergized. If dust accumulation is excessive, deenergize the transformer in accordance with established safety procedures, remove side panels and vacuum away as much dust as possible. Then clean with lint-free rags or soft-bristled brushes. Do not use any solvents or detergents as these may react with the varnishes or insulating materials and lead to accelerated deterioration. They may also leave residues that will enhance future contamination accumulation.

If there is still any dust accumulation in inaccessible areas after vacuuming all loose dust, dry air may be blown in to clear the ducts. Use air or nitrogen that has a dew point of -50°F or less and regulate the pressure at or below 25 psi.

When a transformer is deenergized and the access panels removed for cleaning, all insulation surfaces should be inspected for signs of discoloration, heat damage, or the tree-like patterns etched into the surface that are characteristic of corona damage. The core laminations should be inspected for signs of arcing or overheating. All accessible hardware should be checked for tightness. The isolation dampeners between the base of the transformer and the floor should be checked for deterioration. Any cooling fans or auxiliary devices should be inspected and cleaned at that time also.

If the transformer is deenergized long enough to cool to ambient temperature, care must be taken to assure that the unit is kept dry. If the ambient air is very humid it may be necessary to heat the transformer with electrical strip heaters to avoid the condensation of moisture on the winding insulation. This is a very important point since a large percentage of dry-type transformer failures occur after extended shutdowns when the insulation is allowed to cool and moisture in the ambient air condenses on the insulation.

Pressure readings should be checked and recorded for transformers with sealed tank construction. The ambient temperature, time of day, and loading conditions should be recorded along with the pressure.

The audible sound should be listened to with concentration on the characteristics as well as the sound level. Any noticeable change in the sound level or characteristics should be recorded. Significant changes could be indicative of loose clamping hardware, defective vibration isolators, over excitation, or possibly damage to the primary winding insulation.

If the transformer is in an enclosed space, the effectiveness of the ventilation can be verified by measuring the temperature of the air at the inlet to the room (which should be near the floor) and measuring either the ambient temperature of the air in the enclosed space or the temperature of the air at the exhaust (which should be in the upper part of the room). The average temperature of the room should not increase more than 5°C over the incoming air and the exhaust should not increase more than 15°C. ANSI/IEEE C57.94 gives additional details on ventilation requirements.

Periodic tests. Periodic tests should be performed as often as needed depending on the operating environment. If routine inspections indicate that cleaning is often required, periodic tests should be made following the cleaning operation. The nominal period between scheduled tests is one year but this may be longer or shorter depending on the observed accumulation of contamination on the cooling vents.

Sealed units should be opened only when the need is indicated by loss of pressure, operating abnormalities, or at intervals as recommended in the manufacturer's instructions. On these units periodic tests should be confined to external inspections of the bushings and enclosures and readings of insulation resistance, power factor, and turns ratio taken at the external terminals.

Insulation resistance. Each winding should have its insulation resistance measured with a megohmmeter in accordance with ANSI/IEEE C57.12.91, *IEEE Standard Test Code for Dry-Type Distribution and Power Transformers*, sections 10.9 through 10.9.4. The transformer should be deenergized and electrically isolated with all terminals of each winding shorted together. The windings not being tested should be grounded. The megohmmeter should be applied between each winding and ground (high voltage to ground and low voltage to ground) and between each set of windings (high voltage to low voltage). The megohm values along with the description of the instrument, voltage level, humidity and temperature should be recorded.

The minimum megohm value for a winding should be 200 times the rated voltage of the winding divided by 1000. A winding rated at 13200 volts would thus have a minimum acceptable value of 2640 megohms. If previously recorded readings taken under similar conditions are more than 50% higher, the transformer should be thoroughly inspected and acceptance tests performed before energizing.

Turn-ratio test. The transformer turn ratio is the number of turns in the high voltage winding divided by the number of turns in the low voltage winding. This ratio is also equal to the rated phase voltage of the high voltage winding being measured divided by the rated phase voltage of the low voltage winding being measured. Transformer turns ratio measurements are best made with specialized instruments that include detailed connection and operating instructions. ANSI/IEEE C57.12.91 describes the performance and evaluation of these tests. The measured turns ratio should be within 0.5% of the calculated turns ratio. Ratios outside this limit may be the result of winding damage that has shorted or opened some winding turns and should be investigated.

Insulation power factor. The ratio of the power dissipated in the resistive component of the insulation system when tested under an applied AC voltage divided by the total AC power dissipated defines the insulation power factor. A perfect insulation would have no resistive current and the power factor would be zero. As power factor increases the concern for the integrity of the insulation does also. The power factor of insulation systems of different vintages and manufacturers varies over a wide range (from under 1% to as high as 20%) so it is important to establish a historical record for each transformer and use good judgment in analyzing the data for significant variations. ANSI/IEEE C57.12.91 describes the performance and evaluation of these tests.

Induced voltage tests. ANSI/IEEE C57.94 indicates that induced voltage tests or applied voltage tests may be made as periodic tests. These tests are overvoltage tests and even with proper equipment and skilled operators they are not necessarily nondestructive. They may cause damage to insulation that would otherwise be serviceable at operating voltage level. For that reason they are not considered to be maintenance tests, which are inherently nondestructive.

Acceptance tests. When the need for acceptance tests is indicated, they should include the insulation resistance test, insulation power factor measurements, and turns ratio tests described under periodic tests as well as winding resistance measurements, and excitation current tests. When there is particular cause for concern such as a significant fault in the secondary circuit or a severe overload, an impedance measurement and possibly an applied voltage test may be made.

All these tests have been covered previously in this chapter.

OIL-FILLED TRANSFORMER TESTS

The same basic series of routine checks, periodic tests and maintenance acceptance tests that are used for dry-type transformers also apply to oil-filled transformers. There are, however, others that specific only to these types of transformers.

For oil-filled transformers, the data log should be reviewed for variations in oil temperature, tank pressure or liquid level that cannot be explained by changes in load. In making comparisons, however, remember that the reading from an indicator has a temperature rise added to the oil temperature. This temperature rise varies with the winding load. Therefore, the reading will always be higher than the top oil temperature. The reading should be within 10 to 20°C higher than the oil temperature for a fully loaded transformer.

Internal inspection. Without removing the oil, an internal inspection should be conducted. Open a manhole, handhole, or other cover plate that will allow a view of the internal workings of the transformer. Extreme care must be taken to prevent the entrance of moisture, contamination, or foreign objects. Shine a light through the opening and observe the condition of visible internal components for evidence of internal damage or shifting. The oil should be clear enough to see internal components. Observe steel members above and below the oil level for signs of rust or carbon. Note the oil level and make sure it is above any openings for external cooling equipment. Look for any debris or contaminant on the surface of the oil. Record all observations.

Oil deterioration. Water is the predominant factor in reduced insulation quality of transformer oil. Moisture is everywhere in the atmosphere, and it cannot be totally excluded from transformers. Temperature is the major factor that determines the different quantities of water dissolved into the liquid.

Deterioration of the oil with time also leads to the buildup of acids and such contaminants as varnish, resulting in sludge formation. This reduces the cooling capability and will ultimately result in transformer failure. Deteriorated oil is at or near the sludging stage and is accompanied by low dielectric value. Periodic testing of the oil will show whether its characteristics are adequate and what corrective action must be taken if the quality is found to be poor.

Drawing a sample. All tests begin with the drawing of a sample from the transformer. Simple as that sounds, the results of the tests can sometimes be misleading if some simple precautions are not observed. First of all, transformer tanks filled with warm oil, after cooling, will often have free water accumulated at the bottom. When drawing a sample for testing, this water should be drained off by letting the transformer oil run until it seems to be free of water. If excessive free water is observed when taking a sample it should be determined if the source is a weather leak. If so, this must be corrected before proceeding with further testing.

To prevent false results due to contamination from previous samples, the sample bottle should be clean, and samples should be taken on a clear, low-humidity day. The correct procedure is described in ASTM Standard D-923.

Jobsite tests. The following are tests that can be carried out at the jobsite, using portable test equipment.

Dielectric test, also called a "breakdown test," shows the voltage at which electrical failure occurs in a sample of oil tested under conditions described in ASTM Standard D-877. It also serves to indicate the presence of contaminants and free or suspended (but not dissolved) water. This test is universally made on the oil to determine whether it is in safe operating condition or requires reconditioning.

ASTM standards require that dielectric tests be done at a temperature of 20°C (68°F). The reason for this is that the amount of water that can enter into a solution with the oil depends upon temperature: a higher temperature will reduce the amount of free water; a lower temperature will increase the amount. Either nonstandard condition will produce inaccurate test results.

To carry out the dielectric tests, some of the oil in the sample bottle is poured into the cup of an approved test set. Voltage between the electrodes in the cup is raised at the approximate rate of 3kV/sec until arcing between the electrodes occurs and the unit trips. The kV value indicated on the test unit's meter is them recorded. A minimum of five breakdown tests should be made on a sample in the cup and the readings taken is averaged to determine the dielectric strength of the sample. Breakdown at less than 25kV must be investigated. If the sample is questionable, the oil should be resampled. Certain other circumstances under which the tests may have to be repeated are described in ASTM D-877.

Color of an insulating oil is determined by transmitted light, and is expressed by a numerical value based on comparisons with a series of color standards. A rapidly increasing or high color number is an indication of deterioration, contamination, or both.

This test, per ASTM Standard D-1524, is suitable for gauging color of service-aged electrical insulating liquids.

Lab tests. There are several other tests that are routinely made on samples of oil taken from a transformer.

Dissolved water. Low values of dielectric strength often are an indication of excessive moisture in the sample. Dielectric strength, however, is only a good indicator of *free water* in the oil, but it also may point to excessive *dissolved water*. Laboratory-type equipment is required for this test; therefore, it is seldom used except as a backup to the dielectric test and for other specialized purposes. Water content of 30 ppm or above indicates that drying of the oil may be required, depending upon the transformer operating voltage.

ASTM Standard D-1315 Extraction Method and ASTM Standard D-1533 Karl Fisher Method, are generally used to determine the quantity of dissolved water.

Acidity. This test is used as a gauge of deterioration and sludge buildup. As such, it is a highly valuable tool in the evaluation of field samples and as a control test in reclaiming transformer oil.

In the acidity test, the total acid number is the quantity of the base (expressed in milligrams of potassium hydroxide) that is required to titrate all acid constituents present in one gram of the sample. Good new oil is expected to have no measurable acidity (acid numbers less than 0.02 mg KOH).

The acid neutralization value is determined by color indication titration, done according to ASTM Standard D-974.

Interfacial tension (IFT) is a measure of the concentration of polar compounds (sludging), which may be due to deterioration products, or from contaminants such as varnishes or metallic soaps.

In dealing with a sample of oil from the transformer, a high IFT value is an indication that it is not deteriorated or contaminated with soluble polar compounds. A low interfacial tension indicates that the sample is damaged, but it does not indicate whether the damage is due to deterioration, contamination, or both. Reclaiming is justified if the IFT is below 20 Dynes/cm. Readings below 25 Dynes/cm should be investigated by resampling.

IFT is measured by the drop-weight method as prescribed in ASTM Standard D-2285.

TESTING LESS-FLAMMABLE TRANSFORMER LIQUIDS

Many maintenance persons are familiar with testing procedures for mineral oil used in transformers, but are not sure if the same tests apply to the less-flammable liquids such as silicone (PDMS) and high molecular weight hydrocarbon (HMWH). Indeed, the testing procedures for oil are generally applicable but need some modification in techniques and interpretation.

Besides the need to assure that the sample drawn is clean and uncontaminated, the most important thing to bear in mind is the possibility of cross-contamination. Both of the less-flammable liquids have been used in retrofilling askarel-filled transformers and sometimes even oil-filled ones. Cross-contamination of the fluid can occur due to leaching from the windings, or from the filtering unit and hoses used in retrofilling.

Cross-contamination can result in increased flammability of the transformer fluid. Checking fluid flashpoint and firepoint per ASTM Standard D92 is recommended when cross-contamination is suspected or is likely to occur. Typically, when filtering is required and is to be done with equipment that is not dedicated solely to the type of liquid in use, the flashpoint and firepoint should be checked before and after filtering. Contamination is suspected if the flashpoint is lower than 275°C. A firepoint lower than 300°C is a sign of excessive contamination, and means that the fluid must be replaced or reprocessed.

Silicone. Tests for dielectric strength and water content are the most important ones for PDMS.

Dielectric breakdown strength tests per ASTM Standard D877 are valid for PDMS, but some modification is necessary in the procedures used. PDMS in the cup of a breakdown test set will form a bridge between the electrodes in the presence of an arc. Subsequent test values will not be valid. Thus, unlike oil testing where the test is repeated five times, only one breakdown is made on a sample. The test cup is then drained and the electrodes cleaned with a brush or lintfree tissue, then rinsed with fresh sample liquid. A new sample is poured and tested. The same procedure is followed for a third sample. The dielectric strength of the liquid is taken to be the average of the three values obtained. For maintenance tests of PDMS, the minimum acceptable value is 25kV; lower values indicate excessive moisture of foreign matter.

If the sample has failed the dielectric breakdown test, it is necessary to determine the cause. Water-content tests according to ASTM Standard D1533 (modified) is a logical next step. The modified Karl Fisher method uses dry formamide and dry methanol instead of chloroform and methanol solvent used with mineral oils. Another approach is to use automatic titrators.

PDMS liquids have a higher saturation point than oil, but maintains good dielectric properties with higher levels of water. For this reason, a maximum of 100 ppm of water is considered acceptable.

Color tests on PDMS do not have the same meaning as those for mineral oil. PDMS will experience little if any change in color as it ages. A cloudy liquid may signify the existence of excessive moisture.

Since PDMS liquids are nonpolar and nonoxydizing, tests for acidity and interfacial tension are less useful as indicators of the condition of the fluid than when these tests are used with mineral oil. They may, however, act as an indication of aging and deterioration of winding insulation.

Further directions can be found in ASTM Standard D2225, *Standard Method of Testing Silicone Fluids for Electrical Insulation.*

Paraffinic-based hydrocarbons. HMWH also requires some modification in the tests used for oil and in the interpretation of the results.

Dielectric strength tests remain the basis of determining the condition of the fluid. The procedure is basically similar to oil except that air entrainment is more of a problem. HMWH liquid has a much higher viscosity than mineral oil, therefore, more time must be allowed for air bubbles to escape. Failure to follow this simple procedure will result in misleadingly low test values. While satisfactory results can be attained with HMWH having a dielectric strength of 22kV, values below 26kV are an indication of excessive moisture. The higher value should be considered as the minimum acceptable value of dielectric strength.

HMWH liquid is amber in color and does not change in quite the same way as mineral oil. Aging darkens the color, but unlike conventional oil, this is not an indication of sludge buildup.

Acidity is a useful indicator of the condition of the transformer insulation system. A neutralization number in excess of 0.5 mg KOH per gram is a sign that replacement or reconditioning of the fluid is called for.

Other tests that are used for transformer mineral oil also are applicable to HMWH, but generally are not used.

TRANSFORMER LIQUID RENEWAL

If the cause of low dielectric values can be traced to moisture, the situation can be corrected either by filtering or vacuum dehydration. Here are a few general principles that should be known about these and other processes.

Filtering. The amount of water that can be held in solution by the transformer liquid reduces with the temperature of the liquid. Air holds approximately 20,000 ppm of water at 68°F (20°C) as opposed to 4000 ppm at 32°F (0°C). The maximum amount of water that can be held in solution by liquid at 68°F is 65 ppm, while at 32°F, it drops to 30 ppm. At 176°F (80°C), this figure grows to 220 ppm.

Ideally, oil should be filtered at low temperature to remove the maximum amount of water. Usually this is not practical because the liquid becomes more viscous at lower temperature and, therefore, more difficult to pump.

Vacuum degasification. Filtering only removes free water. To remove soluble water, vacuum degasification makes use of the principal that the boiling point of water is reduced with a decrease in pressure. When a vacuum is pulled on the filtering equipment, it will boil off the water. Experience has shown that if an 80°C core temperature and approximately 15 inches vacuum (½ atmosphere) are maintained, the drying time will be significantly reduced. Water will be removed at the rate of 1 lb per 100 cu ft. After the bulk of the water is removed, vacuum values should be reduced to very low values for thorough drying.

Reclaiming. If other tests show no sludging, filtering may be suitable. If sludging is present, reclaiming or replacement of the liquid is appropriate.

Transformer liquid that has high acidity and/or low IFT can be reclaimed using a large truck-mounted reclaimer. This unit includes a Fullers Earth percolating system, heaters, and a vacuum dehydrator.

Other hints. It is rarely advisable to only filter the "in-service" transformer liquid because if the result of the dielectric test is poor from water only, the winding also is probably wet and the entire unit needs to be dried out. Even if the dielectric strength is good but there is unacceptable acidity or IFT, the liquid should be reclaimed or replaced, whichever is the most economical. For those reasons, filtering should generally only be used while filling or adding liquid to the transformer, or stop-gap maintenance.

Liquid with an IFT in the range of 20 to 25 Dynes/cm range can be used for years without ill effects. At the minimum, however, periodic tests should be made to monitor the IFT and other values. Significant drops in values can signal the need for replacement.

If the dielectric test values indicate a transformer is wet, keep it in service under load and operating near its maximum operating temperature until a drying out program is arranged. The reason is that if it is deenergized and allowed to cool, water will migrate from the liquid to the winding insulation. If the transformer is then again energized, its possibility of failure is much greater.

OTHER INDICATORS

Besides the standard tests of the transformer insulating liquid, there are other indicators of possible problems in liquid-filled transformers.

Oil Temperature. Operating temperatures above normal limits can be indicative of internal problems with the core and coil components or with the normal exchange of heat from the core and coil assembly to the surrounding air. However, the normal operating temperature of a transformer is often the center of a great deal of confusion. Most liquid filled transformers are rated with a temperature rise of either 55 or 65ºC. This rated temperature rise will be printed on the nameplate and is defined in the standards as the average winding temperature rise above ambient. The temperature rise must therefore be added to the ambient or surrounding air temperature to arrive at the expected full load temperature for existing conditions.

Even with the understanding that the nameplate temperature rating is an average winding temperature rise above ambient there is no gauge on a transformer that measures the average winding temperature because it cannot be directly measured. It can only be measured through a series of tests that would be impractical to make outside a factory test installation. The temperatures measured on the gauges are the top oil temperature and sometimes a simulated winding temperature. Both of these measurements will indicate the temperature rise plus the ambient.

The important point to note is that the nameplate temperature rise is not the temperature to expect to see on the gauge when operating at full load. The expected gauge temperatures cannot be precisely derived from the nameplate temperature rise with the information available to most user's but they can be approximated. The relationship between the measured top oil temperature and the average winding temperature varies somewhat from design to design but usually the top oil temperature will be 5 to 10ºC lower than the average winding temperature. The winding temperature indicator, if one is provided, will usually read from 5 to 10ºC higher than the average winding temperature. Remember that all measured temperatures must have the ambient temperature subtracted to derive the temperature rises referred to on the nameplate and in the standards.

Another variable confounding the analyses of temperature readings is the time delay experienced between a change in load or ambient temperature and the eventual transformer temperature change. The time to reach a temperature equilibrium following a change in either or both of these conditions can be 4 hours or more on a typical transformer in a commercial application. It is therefore advantageous to compare temperature readings at the same time of day. If this is done the variables will be minimized assuming that the patterns for load variations and changes in ambient temperature are somewhat consistent for corresponding time periods.

The effect of varying ambient temperature over long time spans can be eliminated in large part if the ambient temperature is recorded so that it may be subtracted from temperature values read on the gauges.

If temperatures under similar load conditions are showing an increase when ambient temperatures are subtracted, there may be thermal problems developing in the transformer and acceptance tests including winding resistance measurements and dissolved gas analyses should be performed and compared with prior tests.

Gas pressure. If the transformer is equipped with a pressure gauge, these readings should be compared for periods of similar top oil temperatures. Whether a transformer has a pressure gauge or not depends on the type of oil preservation system. The general types of oil preservation systems are: free breathing, sealed, conservator, and automatic gas seal, with variations of each of these types. Most transformers in commercial applications are either free breathing or sealed. A sealed unit will usually have a pressure gauge, a free breathing unit will not. A free breathing unit has vents above the oil that allow air to enter and exit as the oil expands and contracts due to variations in the operating temperature. A sealed unit does not have vents but is designed to withstand the internal pressure variations resulting from the compression of the gas space above the oil as the oil volume changes due to thermal expansion and contraction.

A sealed transformer with a welded on cover should maintain a consistent relationship between top oil temperature and pressure. If a review of the maintenance record indicates that periods of maximum temperature do not have correspondingly high pressure readings with minimum pressure readings at lowest temperatures, a leak in the gas space should be suspected. If the peaks and valleys of the pressure readings correspond to similar peaks and valleys in the temperature readings but the values of the pressure readings are declining over time, check the liquid level for loss of fluid and inspect the transformer for fluid leaks.

Many sealed units with bolted on covers and gasket seals will lose gas pressure if a positive pressure is maintained for an extended period of time. These same units may allow the entrance of air if a negative pressure is maintained over an extended period. The concern for this condition should depend on the ambient weather conditions (humidity, precipitation, and airborne contamination), and the degree of cyclic variations in liquid temperature that might tend to create conditions that will draw moisture or other contaminants into the transformer through a leak.

When conditions exist that would tend to allow the entrance of contaminants and the pressure readings indicate a leak, the transformer should be deenergized and a pressure test performed to detect and evaluate any possible leaks. Most leaks can be found and effectively sealed but large gasket areas especially those using cork or composition gaskets will often allow the gradual diminishment of gas pressure even though an identifiable leak cannot be found.

On operating transformers with minor gas leaks, tests of the fluid dielectric and water content should be monitored closely and if there is no noticeable deterioration the concern for a gradual loss of gas pressure should be minimal.

Liquid Level. The liquid level should vary with the top oil temperature just as the tank pressure on a sealed transformer does. If the indicated liquid level pattern does not follow the rise and fall of the top oil temperature, investigate for oil leaks. If no oil leaks can be found, the operation of the liquid level gauge should be checked at a convenient outage.

Liquid level should not descend below the minimum indication on the gauge or rise above the maximum indication in service extremes. If these limits are exceeded, the manufacturer or instruction book should be consulted to establish the proper oil level and the existing level checked at the earliest available outage. If the oil level is consistently below the minimum indication, the transformer should not be operated until the internal level is checked to ensure that no live parts are exposed and the minimum oil level reaches the upper tank opening of any existing external cooling radiators.

The manufacturers' instructions should be followed for adding any oil to the transformer. If the instruction book cannot be located, the manufacturer should be contacted. There are often critical variations in the replenishment fluids used and the manner in which they are introduced to the fluid already in the transformer.

Sound Level. The audible sound level of a transformer is largely dependent on the ratio of the applied voltage to the number of active turns in the primary winding (volts per turn) or on the degree of distortion in the load current. It is dependent to a lesser extent on the tightness of core and coil clamping components and the external tank structure.

If a noticeable change in sound level is detected that cannot be explained by changes in loading practices, the first check that should be made is to measure the input or output voltage on the transformer. The sound level of a transformer is very sensitive to changes in voltage. If the voltage increases, the sound level will also increase. Check to see if the measured voltage is within the nameplate rating for the tap setting on the transformer. If the measured voltage consistently exceeds the tap voltage by more than 5%, the transformer is overexcited and should be deenergized and a tap selected that is within 5% of the applied voltage. Transformers designed to existing standards can be safely operated at overvoltages of up to 5% but the sound level will increase noticeably.

If the applied voltage is within the range of the tap setting on the transformer and there is an unexplained increase in sound level, there could be internal damage that has shorted one or more winding turns in the primary winding. This would reduce the effective number of turns and increase the volts per turn and the sound level. The transformer should be removed from service for acceptance tests. If these are inconclusive and the transformer is to remain in service, samples should be taken for gas-in-liquid analysis on at least a monthly basis until their analyses refute or confirm the likelihood of internal winding problems.

An increase in sound level can also be the result of a distorted load current. Check the connected load for any changes. There may possibly be a developing problem with a load component that has introduced load distortion on the transformer. As various load segments are switched in and out listen to the sound level for any abrupt changes.

Tank heating. Hot spots on the tank surfaces that are severe enough to blister or discolor the paint may indicate the existence of open or shorted internal lead connections. These deficiencies may create changes in the current paths that induce currents in the tank wall. The transformer should be deenergized as soon as possible and electrical tests performed. Winding resistance and impedance measurements are especially important when tank heating is observed because changes in these characteristics would indicate changes in the internal connections. An sample should be taken for gas-in-liquid analysis.

An internal inspection should be made to observe any evidence of irregularities in the internal connections. The surface of the liquid should be examined for evidence of carbon or burnt insulation. If the liquid is discolored to the point that the internal parts cannot be seen, the transformer should be removed to a repair facility for untanking and examination.

If a defective connection is identified, an experienced repair organization may be able to repair and recondition the transformer for reliable service.